Biosystems & Biorobotics

Volume 8

Aims & Scope

Biosystems & Biorobotics publishes the latest research developments in three main areas: 1) understanding biological systems from a bioengineering point of view, i.e. the study of biosystems by exploiting engineering methods and tools to unveil their functioning principles and unrivalled performance; 2) design and development of biologically inspired machines and systems to be used for different purposes and in a variety of application contexts. The series welcomes contributions on novel design approaches, methods and tools as well as case studies on specific bioinspired systems; 3) design and developments of nano-, micro-, macrodevices and systems for biomedical applications, i.e. technologies that can improve modern healthcare and welfare by enabling novel solutions for prevention, diagnosis, surgery, prosthetics, rehabilitation and independent living.

On one side, the series focuses on recent methods and technologies which allow multiscale, multi-physics, high-resolution analysis and modeling of biological systems. A special emphasis on this side is given to the use of mechatronic and robotic systems as a tool for basic research in biology. On the other side, the series authoritatively reports on current theoretical and experimental challenges and developments related to the "biomechatronic" design of novel biorobotic machines. A special emphasis on this side is given to human-machine interaction and interfacing, and also to the ethical and social implications of this emerging research area, as key challenges for the acceptability and sustainability of biorobotics technology.

The main target of the series are engineers interested in biology and medicine, and specifically bioengineers and biorobotics. Volume published in the series comprise monographs, edited volumes, lecture notes, as well as selected conference proceedings and PhD theses. The series also publishes books purposely devoted to support education in bioengineering, biomedical engineering, biomechatronics and biorobotics at graduate and post-graduate levels.

About the Cover

The cover of the book series Biosystems & Biorobotics features a robotic hand prosthesis. This looks like a natural hand and is ready to be implanted on a human amputee to help them recover their physical capabilities. This picture was chosen to represent a variety of concepts and disciplines: from the understanding of biological systems to biomechatronics, bioinspiration and biomimetics; and from the concept of human-robot and human-machine interaction to the use of robots and, more generally, of engineering techniques for biological research and in healthcare. The picture also points to the social impact of bioengineering research and to its potential for improving human health and the quality of life of all individuals, including those with special needs. The picture was taken during the LIFEHAND experimental trials run at Università Campus Bio-Medico of Rome (Italy) in 2008. The LIFEHAND project tested the ability of an amputee patient to control the Cyberhand, a robotic prosthesis developed at Scuola Superiore Sant'Anna in Pisa (Italy), using the tf-LIFE electrodes developed at the Fraunhofer Institute for Biomedical Engineering (IBMT, Germany), which were implanted in the patient's arm. The implanted tf-LIFE electrodes were shown to enable bidirectional communication (from brain to hand and vice versa) between the brain and the Cyberhand. As a result, the patient was able to control complex movements of the prosthesis, while receiving sensory feedback in the form of direct neurostimulation. For more information please visit http://www.biorobotics.it or contact the Series Editor.

More information about this series at http://www.springer.com/series/10421

Francisco J. Valero-Cuevas

Fundamentals
of Neuromechanics

 Springer

Francisco J. Valero-Cuevas
Department of Biomedical Engineering
The University of Southern California
Los Angeles, CA
USA

and

Division of Biokinesiology and Physical
  Therapy
The University of Southern California
Los Angeles, CA
USA

Additional material to this book can be downloaded from http://extras.springer.com.

ISSN 2195-3562   ISSN 2195-3570   (electronic)
Biosystems & Biorobotics
ISBN 978-1-4471-7089-1   ISBN 978-1-4471-6747-1   (eBook)
DOI 10.1007/978-1-4471-6747-1

Springer-Verlag London Ltd. is part of Springer Science+Business Media (www.springer.com)

Philosophy [nature] is written
in that great book which ever is before
our eyes—I mean the universe—but
we cannot understand it if we do not first
learn the language and grasp
the symbols in which it is written.
The book is written in mathematical
language, and the symbols are triangles,
circles and other geometrical figures,
without whose help it is impossible
to comprehend a single word of it;
without which one wanders in vain
through a dark labyrinth.

—Galileo Galilei, 1564–1642

*To my parents Enrique and Margarita,
and to Erika, Marco, and Eva*

Foreword by An

During the last half-century, the fields of Biomechanics and Motor Control each experienced significant development and made important contributions to movement science and rehabilitation medicine. Unfortunately, scientists and students from these two fields did not fully appreciate knowledge from the other field. There was little collaboration between them, and in fact, there were even misconceptions in one field about the other. This book, *Fundamentals of Neuromechanics*, carefully prepared by Francisco, provides a single go-to reference for those wishing for a better understanding of each of the fields, offering its reader an appreciation of muscle function in human movement. It is written using basic mathematical language and introduces the foundations of rigid body mechanics and neuroscience at a readily accessible level. I expect this text will facilitate collaborations between the fields, which are likely to be fruitful and yield further progress in the combined fields. After reading this book, I was reminded of *Fearfully and Wonderfully Made*, by Philip Yancey and Dr. Paul Brand. Like the universe, our body and its neuro-musculoskeletal system were fearfully and wonderfully designed and created by God our Creator. Our inquisitive nature and God-given scientific tools enable us to in turn discover His Creation.

Rochester, MN, USA

Kai-Nan An
Professor, Mayo Clinic College of Medicine

Foreword by Biewener

Drawing from his own work on the motor control and biomechanics of hand and finger manipulation and gripping tasks, Prof. Valero-Cuevas admirably assembles a valuable synthesis of essential kinematics and mechanics concepts drawn from engineering, which he links to neuromuscular function and engagingly applies to current problems of motor control. In doing so, the author compellingly argues that muscle redundancy or indeterminacy may not be the central problem it has long been considered in motor control.

Instead, it likely facilitates the control of more complex movement dynamics. The resulting integration will provide a much needed and key resource for the neuromechanics community interested in addressing problems that range from the control of human and animal movement to robotics and rehabilitation engineering.

July 2015

Andrew A. Biewener
Charles P. Lyman Professor of Biology
Harvard University

Foreword by Giszter

In this book Francisco Valero-Cuevas has provided an excellent reference and a powerful teaching tool. He provides a very thorough set of tools and materials which cover and summarize crucial introductory concepts that any student of neuromechanics must learn in order to understand the current literature in this area, and gain a firm 'hands on' grasp of the subject in order to take a deeper critical view. The book is well thought out and organized, and is an excellent addition to the motor literature and course texts that are available. It fills a much needed gap, and very nicely augments and complements the information in other recent books in motor control such as *Human Robotics*, *The Computational Neurobiology of Reaching and Pointing*, and *Biological Learning and Control*. The graduate student of Motor Control today has available a bookshelf that was sorely absent at the time I was a graduate student and postdoc.

Francisco's book is an essential addition to the Motor Control bookshelf, and encapsulates his personal perspectives and areas of intensive research thoroughly, while educating the reader in the broader range of ways that motor data can be considered. As a teacher of a Motor Control class myself, teaching engineers, neuroengineers, neuroscientists, and physical therapists and bringing them to a point of common understanding and discourse, this book is a very welcome text, and one that will be well used by my students.

In summary, this book should be among the resources on the bookshelf of every researcher and student of Motor Control.

June 2015   Simon Giszter
Professor, Neurobiology and Anatomy Drexel University College
of Medicine and Joint Faculty in School of Biomedical Engineering
and Health Systems, Drexel University, Philadelphia, PA, USA

Foreword by Milton

It is often observed that macroscopic behaviors are governed by much simpler rules than would be anticipated from detailed descriptions of microscopic interactions. Well-studied examples of the applications of simple mathematical rules to biology range from the synchronization of oscillators to the self-organized behaviors or schools of fish, flocks of birds, and herds of animals.

In this book *Fundamentals of Neuromechanics*, Professor Valero-Cuevas provides another example of the principle that "simplicity explains complexity" through his study of vertebrate limb function. Contractile proteins can only make muscles actively pull on their tendons which attach to bones after crossing joints. Consequently limb function in vertebrates is not torque-driven as in the mathematics of robotics, but tendon-driven. Thus many of the mathematical problems associated with the analysis of limb function can be elegantly reduced to considerations of the addition and subtraction of vectors.

Professor Valero-Cuevas provides a powerful demonstration that mathematics is a necessary laboratory tool for biologists. The often-overlooked observation that the mechanical function of limbs is tendon-driven means that linear algebra can be used to understand a number of otherwise perplexing aspects of the neural control of limb function. It is remarkable that so much can be understood about the neural control of limb function using only linear algebra and principles typically taught in introductory courses in mathematics and mechanics.

The fact that the predictions can be readily verified even in simple simulations and in the teaching laboratory will motivate the development of inclusive courses to illustrate the multidisciplinary nature of scientific discovery.

It is not difficult to imagine that this textbook could be used as the backbone for a team-taught course on neuromechanics which combines mathematics, biomechanics, dissection of human cadavers, and measurements of limb function. Such a course would address a fundamental challenge for STEM (science, technology, engineering, mathematics) educational initiatives, namely the development of inclusive and interdisciplinary courses. Thus it would be possible to highlight the multidisciplinary nature of scientific discovery in a manner that attracts interest from students with a wide variety of backgrounds.

June 2015   John Milton
                                           The Claremont Colleges, Claremont, CA, USA

Preface

This book is born from the need for a succinct means to convey basic concepts of engineering, mechanics, and mathematics to those who care deeply about combining neuroscience and biomechanics to forge an understanding of the function, control, and rehabilitation of vertebrate limbs. And conversely, to convey my fascination with organisms and neuroscience to engineers who are inspired by Nature to create versatile robots. The principles of mechanics are at the root of both evolutionary biology and robotics. Darwinian evolution and Newtonian mechanics are unforgiving arbiters that continually shape what is possible and successful in the physical world. Thus animals have had to evolve to successfully withstand and exploit the laws of mechanics.

The scope of this book on neuromechanics is heavily biased toward mechanics because it is least emphasized in today's efforts in neuroscience. We have historically emphasized reductionist neurophysiological preparations and recordings that have given us such triumphs as the Hodgkin–Huxley model that describes how action potentials in neurons are initiated and propagated, as well as the understanding of neuron structure, function, and connectivity—and their role in memory and learning—in the tradition of Ramón y Cajal, Brodmann, and Kandel, to name a few. At the larger scale of populations of neurons, we have the discovery and description of central pattern generators, the connectome, and cortical networks. And medical imaging now even allows us to obtain whole-brain views in real time with MRI, fMRI, PET, etc.

In parallel, we have the emergent understanding of whole-body function from the biomechanical perspective, as first introduced by Borelli and Da Vinci, which is now a mature field able to study complex systems using computer aided design platforms such as openSIMM, AnyBody, and MSMS. Similarly, muscle physiology, structure, and function is an active field of study that has enabled our ability to understand the nanomechanical interactions and molecular motors at the center of sarcomere and muscle function. Add to this the crucial contributions from electrical engineering and signal processing that have allowed us to analyze physiological signals, such as the electromyogram and electroencephalogram, to extract estimates of motor commands and brain function, respectively.

But, in the domain of understanding the principles, strategies, and mechanisms that the nervous system uses to control vertebrate limbs, we have theories that remain biased toward the notion that muscle redundancy is the central problem of motor control. Thus, the problem that the brain solves is cast as one of mathematical optimization. However, this classical notion of muscle redundancy is paradoxical with respect to the evolutionary process and clinical reality.

This book at its core seeks to reconcile these conflicting views: engineers and neuroscientists who think that we have too many muscles, versus evolutionary biologists and clinicians who cherish every one of them as functionally relevant. This motivates questions of why vertebrates have evolved the way they have, and what neural computational problems confront the brain when it controls its mechanical body. My hope is that the presentation of this mechanically-centered material will complement current thinking and advance the state of the art in robotics, neuroscience, and rehabilitation.

I write this book with trepidation as I inevitably run the risk of potentially—yet without intent—antagonizing esteemed colleagues by acts of omission or over-simplification of the many nuanced issues and concepts that have critical bearing on this topic. But this is an unavoidable risk, as it is intended as a succinct introductory textbook for upper class or early graduate students and working scientists. I also provide extensive online supplemental material, references, and suggested reading to help fill in the inevitable technical gaps and scientific nuances.

Nevertheless, I believe that engineering is critical to progress in life sciences—and that this particular neuromechanical approach is underdeveloped and has lacked sufficient attention. My experience over the years of teaching related courses at Cornell University and the University of Southern California confirms my belief that it enables engineers, neuroscientists, and clinicians to evolve their intuition and understanding of neuromusculoskeletal systems. It is my sincere hope that it will do the same for you.

Los Ángeles, CA, USA   Francisco J. Valero-Cuevas
June 2015

Acknowledgments

This book is the direct result of the opportunity and privilege I have had of working with passionate and skillful mentors, colleagues, and students. I cannot give due credit to all, but some deserve special mention.

Fredrick Orthlieb, Faruq Siddiqui, Rachel Mertz, and Brian Clark actively promoted and guided my efforts to combine engineering with biology in my undergraduate training at Swarthmore College. My late mentor Carolyn Small demanded and encouraged intellectual independence from me at Queen's University. My doctoral mentors Felix Zajac, Rod Hentz, and Chuck Burgar gave me every opportunity while holding me to the highest standards at Stanford University. Art Kuo and Kai-Nan An were also instrumental to my development during my doctoral work.

At Cornell University, John Guckenheimer, Hod Lipson, and Andy Ruina were —and continue to be—incomparable colleagues and friends. Carlos Bustamante and my students Madhusudhan Venkadesan, Veronica Santos, Kevin Keenan, Jon Pearlman, Robert Clewley, Roberto Malvaez, and Stanley Song, among others, helped bring the material in this book into focus.

At the University of Southern California, Manish Kurse, Jason Kutch, Kornelius Rácz, Sarine Babikian, Cassie Borish, Brian Cohn, Josh Inouye, and Emily Lawrence contributed in ways big and small to the execution of this book, often contributing computational examples, figures, and ideas scattered across several chapters.

Komei Fukuda at ETH-Zürich deserves special mention for freely giving of his time and expertise to make the computational geometry aspects of my doctoral dissertation possible, and also for actively promoting a collaboration with Bernd Gärtner, Hörður Yngvason, and May Szedlak. These last two contributed actively to this book.

Emo Todorov, Jerry Loeb, Simon Giszter, Andrew Biewener, Lena Ting, Konrad Körding, Raff D'Andrea, John Milton, Bob Full, Walter Herzog, Sandro Mussa-Ivaldi, Andrea D'Avella, and Rodger Kram, among many others, have been constant companions in my exploration of these areas during my academic career.

I gratefully acknowledge the research support related to the topic of this book from the Whitaker Foundation, the Swiss National Science Foundation, US National Science Foundation (NSF) grants CAREER BES-0237258, IGERT 0333366, and EFRI-COPN 0836042, as well as grants R01-AR050520 and R01-AR052345 from NIAMS of the US National Institutes of Health (NIH). The contents are solely the responsibility of the author and do not necessarily represent the official views of the NSF or the NIH.

Contents

Chapter 1
Introduction

Abstract Neuromechanics is a perspective that highlights how real-world behavior emerges from the intimate relationship between the mechanical structure of the musculoskeletal system, the mechanical requirements of a task, and the feasible neural control actions to produce it. To understand these interactions, it is necessary to consider the anatomical fact that muscles act on vertebrate limbs via tendons. This is different from the more common mathematical formulation that focuses on analyzing the net action of all muscles at each joint. This deliberate consideration of tendon-driven limbs allows us to articulate the problem of neural control in a way that promotes the debate and refinement of current theories. This perspective has important consequences to understanding healthy function, disability, and rehabilitation; and to the design of novel versatile robots.

Motivation and Goals

Understanding the interactions between the brain and body is critical to understanding neural computation, vertebrate evolution and function, clinical dysfunction and rehabilitation—and to creating truly versatile robots. In recent decades, this scientific endeavor has been dominated by the perspective that muscle redundancy (i.e., the fact that we have 'too many' muscles) is the central problem of motor control.[1] Therefore the 'problem' the brain solves is cast explicitly and/or implicitly as one of neural computation to select specific commands to muscles from the many (infinite, in fact) options allowed by that redundancy. However valuable and informative this perspective has been, it is also paradoxical with respect to the evolutionary process and clinical reality. That is:

- Why would organisms evolve, encode, grow, maintain, repair, and control unnecessarily many muscles when a simpler musculoskeletal system would suffice, and thus have phenotypical and metabolic advantages?
- Why would musculoskeletal systems evolve in such a way as to require the nervous system to solve an optimization problem with infinite solutions?
- Why do people seek clinical treatment for measurable dysfunction even after injury to a few muscles, or mild neuropathology?

---

[1] 'Motor Control' is the monicker that the community of neuroscientists, roboticists and clinicians use to refer to the neural control of the musculoskeletal system. I see and use the term as synonymous with *neural control*, *neuromuscular control*, and *sensorimotor control*.

© Springer-Verlag London 2016
F.J. Valero-Cuevas, *Fundamentals of Neuromechanics*,
Biosystems & Biorobotics 8, DOI 10.1007/978-1-4471-6747-1_1

1

- Which muscle would you donate to improve your neural control?
- Why is it that, despite our best efforts, engineers are still far from creating robots with versatility comparable to that of vertebrates?
- How can versatile function arise from muscles that have such nonlinear viscoelastic properties, and sensors that are so delayed and noisy?

This book is intended for neuroscientists, roboticists, engineers, physicians, evolutionary biologists, physical and occupational therapists, athletes, etc. seeking to explore these apparent paradoxes. While no one approach can hope to resolve these important questions, I provide background in the context of vertebrate limbs to begin to create a unified approach to these questions. This approach, which highlights the intimate relationship between the mechanical structure of the musculoskeletal system and the feasible neural control actions that can meet the mechanical requirements of real-world ecological behavior, can be called *neuromechanics*.[2]

I understand neuromechanics as a perspective that allows us to view and understand the structure and function of the nervous system from the perspective of the mechanical environment in which it evolved. Likewise, it allows us to understand anatomical structure and function from the perspective of the control capabilities of the nervous system. Indeed, 'neuro'-'mechanics' is a direct result of 'brain'-'body' coevolution for the purposes of producing versatile mechanical function. Such a unifying perspective begins to address the questions listed above.

My goal is to provide a rigorous mathematical and biomechanical foundation to help understand the neuromechanical interactions in vertebrate limbs. The reason for a dedicated book in a field that already has extensive literature is that the elements of this mathematical and biological foundation come from a variety of fields that are seldom taught together, and which have very different cultures, value systems, and terminologies. Neuromechanics of vertebrates includes fundamental concepts from areas such as mechanics, neuroscience, physics, computational geometry, muscle mechanics, anatomy, evolutionary biology, robotics, applied mathematics, and computer science. However, I have made every effort for this text to require the reader to have only a basic knowledge of anatomy, biology, physics, and a modest background in mathematics at the level of linear algebra. I then attempt to provide the necessary concepts, background, and references to give the reader the engineering and analytical tools to become skillful in, and gain insight into, neuromechanics.

The working hypothesis of this text in the context of motor control is that evolutionary pressures drive vertebrates to have, in fact, barely enough muscles to meet the mechanical requirements of ecological tasks. That is, each muscle contributes to function in unique ways that justify its retention in the evolutionary process. The control options open to the nervous system, while numerous for simpler laboratory tasks, become highly structured or even very reduced for real-world ecological tasks. Oversimplifying to make a point, the control problem the nervous system faces would then be more one of pattern recognition, learning, and memory than one of intensive, real-time neural computation and optimization to select from among infinite options.

---

[2]To my knowledge, the term *neuromechanics* was first coined by Roger M. Enoka in his 1988 book Neuromechanical Basis of Kinesiology.

A deceivingly simple idea lies at the core of this presentation of neuromechanics: That to understand motor control we must embrace the fact that muscles act on the body via tendons, i.e., that vertebrate bodies are *tendon-driven* systems. While this has been obvious since the very first anatomical studies, most engineering, neuroscience, biomechanical, and mathematical analyses have tended to simplify the phrasing of the problem of neural control—for multiple good conceptual, analytical and computational reasons. This simplified phrasing of the problem relates mostly to the fact that the mathematics of robotics is best developed for *torque-driven* limbs. In those robotic limbs there are rotational motors at each hinge joint that produce a torque, angle, or angular velocity directly.

Another common simplification is the functional grouping of muscles into fewer control entities approximating the number of anatomical joints (i.e., kinematic degrees of freedom, or DOFs) to allow close-formed solutions or more tractable optimizations. Such approximations have been very valuable to motor control, and instrumental to our understanding of vertebrate function.

The neuroscience and clinical communities have undoubtedly benefitted from these simplifications of how actuators (i.e., muscles) act upon serial linkage systems (i.e., limbs) to produce mechanical function (i.e., motions, forces, and their combinations). But we must keep in mind that such analyses and design tools were developed for systems that are, by definition, not fully compatible with the actual anatomical structure of vertebrates. In my mind, these simplifications and approximations have, in part, led us to the paradoxes listed above.

Alfred Korzybski's dictum applies well here: 'The map is not the territory.' Torque-driven approximations apply well to systems with simple hinges as the kinematic DOFs, with few muscles acting on each DOF. However, researchers working to understand the actual neuromuscular control problem that confronts the brain have no such guarantees, and biological systems often violate fundamental assumptions of those mathematical formulations. In comparison, the mathematics of tendon-driven limbs is less well developed, is less accessible to non-engineers, requires specialized computational methods, and therefore is not as commonly used by the motor control community.

What forced me to confront the nature of tendon-driven limbs? My initial interests in motor control were focused on the neural control of finger function for basic science and clinical applications. Several factors inherent to this area of research conspired to make me pursue a tendon-driven approach. First, fingers have relatively 'few' muscles compared to the dozens of muscles in the limbs, making it attractive to want to create mathematical models that include 'all' muscles. Second, working in the laboratory of my mentor Felix Zajac, I witnessed many heated discussions about the nature and control of bi-articular muscles of the arms and legs (i.e., muscles that act on 2 DOFs)—and could not help but wish I were working with such systems. By any count, all muscles of the fingers are at least bi-articular, with most being tetra-, penta-, and even hepta-articular. Third, the close collaboration with my clinical mentor Rod Hentz—a gifted hand surgeon—instilled in me the healthy attitude of insisting on anatomically and physiologically faithful mathematical models. After all, for surgeons to use our models, those models need to be as realistic as possible.

As such, I could not afford to work with simplified musculature or torque-driven approximations. Thus I devoted my attention to learning what roboticists had developed for tendon-driven robots, and developing a mathematical framework suitable to the analysis of tendon-driven limbs.

Working in collaboration with gifted colleagues and excellent students, first at Cornell University and then at the University of Southern California, we developed advances and formulations of the mathematics of tendon-driven limbs in the context of anatomy, neuroscience, robotics, and motor control. It then became increasingly clear that these neuromechanical properties were common to all tendon-driven limbs—not just fingers or human anatomy, but that they were an inherent feature of vertebrate motor control.

This book's content and structure is designed to present the reader with the key concepts of neuromechanics to develop a nuanced treatment of tendon-driven limbs by confronting the mathematically inconvenient details of vertebrate limbs. This nuanced understanding of tendon-driven limbs then allows one to articulate the problem of motor control in a way that begins to resolve those apparent paradoxes and clarifies key concepts of motor control. Most importantly, it allows a clearer definition of the necessary mechanical requirements for limb function and control (i.e., the problem the brain faces) essential to debate, evaluate, discard, and refine current theories of motor control, disability, and rehabilitation.

Another reason to confront the tendon-driven mathematical structure of vertebrate limbs is that it may hold important lessons to revise and re-direct our approach to robotic limb design. We engineers continue to be puzzled by the fact that—even though the anatomical structure of vertebrates seems to violate some of our preferred engineering design principles—vertebrate limbs often outperform robotic systems in many functional domains. Thus, for example, engineers often prefer not to build robots with multiple or bi-articular tendons, loose joints, sluggish actuators, or compliant tendons with complex routing because they would complicate their control. This is not to say that roboticists have not worked actively on the problem of tendon-driven limbs and complex robots. They have, and have made great progress from which we benefit. However, that work on tendon-driven limbs is often not well known by the community studying neuromuscular control of limbs. But the fact remains that the structure of vertebrates flaunts such considerations and still manages to outperform robots that have the benefit of our sophisticated algorithms, fast computers, and precise sensors and actuators. The material in this book helps explore why this may be the case.

Last, and to underscore the importance of biological and anatomical concepts, it is important to mention how the language used to describe and discuss anatomical systems has in fact confused important engineering and clinical issues, and hindered our progress. For example, we inherited the naming of muscles as 'flexors,' 'extensors,' 'abductors,' 'adductors,' etc., largely from 15th century anatomists, who used these terms to describe the joint rotations induced in a limp cadaver when tendons were pulled in isolation. Over time, however, these names gradually were taken by both biologists and engineers to represent their *functional* role during natural, whole-limb, coordinated action for the production of both motions and forces. The analyses in this book show that such nomenclature is often misleading and, worse, has had the unfortunate consequence of giving us a false sense of certainty about what each muscle does, or how muscles interact. For example, this book addresses terms and concepts like 'agonist,' 'antagonist,' and 'co-contraction;' and shows that they lose their intended meaning and usefulness when relating to realistic limb structure and function. In addition, it is very difficult to find a task, simple as it may seem, where there is a muscle that can be considered a 'prime mover.'

More subtly, and perhaps more damaging to our conceptual progress, is that anatomical nomenclature has led to the propagation of a vague notion—confusion, really—about the role of muscles during the production of limb motion versus static force. As engineers we know that this distinction is so fundamental and clear that it hardly seems worth mentioning: the mathematical formulations of the equations of motion versus the equations for static equilibrium are different because their physics are different. However, this distinction is often lost in the motor control literature and thinking. Therefore, I take this opportunity to underscore these differences as needed, which have profound consequences to our understanding of how tendons need to be controlled to produce isolated free limb motion versus isometric force, and their combination.

It is my sincere hope that this—inevitably imperfect and incomplete—interdisciplinary presentation of biological, neural, analytical, and computational approaches to tendon-driven limbs will help future students define and explain the actual mechanical problem that the nervous system faces to produce versatile function.

Part I
Fundamentals

Neuromechanics requires understanding the fundamental concepts that describe how neuromuscular systems produce mechanical function. This Part presents these concepts in a succinct way. It will allow the reader to define and analyze the way in which multiple muscles act on multiple joints. Appendix A presents a brief overview of the few fundamental concepts of linear algebra and kinematics of rigid bodies used in this book.

Part I
Fundamentals

Chapter 2
Limb Kinematics

Abstract The purpose of this chapter is to introduce you to the *kinematics* of limbs. Kinematics is the study of movements without regard to the forces and torques that produce them. In essence, it is the fundamental description of the articulations and motions of which a limb is capable. This chapter serves as the foundation upon which we can build a common conceptual language, and begin to discuss limb function in the context of mechanics.

2.1 What Is a Limb?

A clear understanding of the kinematics of limbs is necessary to compare and contrast the capabilities and limitations of biological and robotic limbs. Kinematics is the study of the motions and positions of rigid bodies, like limbs, *without* regard to the forces that produce them. The kinematic *degrees of freedom* (DOFs) are the articulations between the links. These articulations (i.e., anatomical joints) are the mechanical structures that allow for changes in the configuration of the links with respect to each other. In general, there are two kinds of engineered DOFs that are convenient to define and use mathematically: the linear or *prismatic DOF* like telescoping tubes that change the lengths of a link in the limb; and the revolute or *rotational DOF* like a hinge that changes the orientation of adjacent links in the limb. This allows us to use the state of each DOF to define a specific limb configuration, shape, and size. In addition, by having motors act on each DOF using linear or rotational motors, we can produce specific limb forces and accelerations.

I will focus on rotational joints because most vertebrate limbs are approximated as behaving in this way,[1] as in Fig. 2.1. Universal joints, such as those used to represent 2 DOF rotational joints like the metacarpophalangeal (MCP) joint of the index finger (which is the base knuckle of the finger), consist of two pin joints with intersecting and perpendicular rotational axes. Ball-and-socket joints, like the shoulder or hip, consist of three intersecting and perpendicular rotational joints. Other joints like the

---

[1] In biological systems, joint kinematics arise from the interaction of the contact of bony articulating surfaces held by ligamentous structures. A joint is, therefore, a complex system whose kinematics can be load dependent [1].

© Springer-Verlag London 2016

F.J. Valero-Cuevas, *Fundamentals of Neuromechanics*,
Biosystems & Biorobotics 8, DOI 10.1007/978-1-4471-6747-1_2

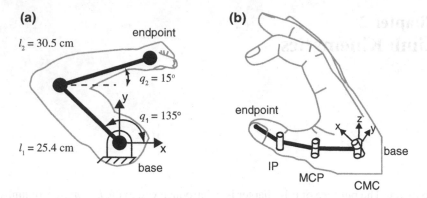

Fig. 2.1 a Human arm modeled as planar 2 DOF serial manipulator. The angle of each DOF is often described as the q_i variable, where i is the index of the DOF. **b** Human thumb modeled as 3D serial manipulator with 5 rotational DOFs contained in the carpometacarpal (CMC with 2 DOFs), metacarpophalangeal (MCP with 2 DOFs), and interphalangeal (IP with 1 DOF) joints. Note the axes of rotation of the different DOFs need not be parallel or perpendicular to each other [10]

knee or jaw have more complex kinematics that involve both rotation and sliding, but they are often approximated as pure rotational joints [2–5].

Note that in this book it is necessary to alternate between anatomical and robotics terminology. Robotic limbs were inspired by vertebrate limbs, and vertebrate limbs are analyzed using the mathematics developed for robotic limbs.

An important distinction between engineered and vertebrate limbs is that the former are often *torque-driven limbs*[2] where motors act on each joint to rotate them, whereas the latter are *tendon-driven limbs* where muscles pull on tendons that act on joints by spanning 1 or more DOFs. The mathematics and theory of torque-driven limbs is well developed, and serves as a starting point to discuss tendon-driven limbs. Therefore, this chapter gives a brief overview of the theory behind the basic kinematic analysis of torque-driven *robotic manipulators*.[3] That is, robotic limbs that have motors actuating on rotational joints. Subsequent chapters leverage this fundamental understanding to discuss the case of tendon-driven limbs. This is important because tendon-driven limbs can be, and often are, simplified into their mathematically equivalent torque-driven limbs—but with important conceptual consequences as mentioned in Chap. 1. This chapter is intended to give the reader a sense of the mathematical principles of these topics and some associated issues, but is by nomeans

---

[2] We use the term torque-driven instead of joint-driven because this is more common in the robotics literature.

[3] In robotics, the term *manipulator* is used synonymously with *robot*, *robotic arm*, *robotic limb*, or any other mechanism that is actuated and controlled.

comprehensive and further study is likely required to use this knowledge in detail. Specialized monographs [6, 7] and our prior publications [5, 8] contain a wealth of detailed information and background material.

The heart of the matter in many debates in neuromechanics is that much of the theory of robotics analysis and design takes a torque-driven approach. And rightfully so as it grows historically out of the mathematical description of serial linkage systems with pin-joints (rotational joints with 1 DOF), universal joints (with 2 DOFs), or ball-and-socket joints (with 3 DOFs), driven by idealized rotational *actuators*.[4] That is, whatever kind of electric, pneumatic, hydraulic, etc. linear or rotational actuator is used in practice (via gears, pulleys, capstans, belts, direct-drive, etc.), it can be analyzed as inducing torques at these rotational joints.break However, some subtle but important points arise when tendon-driven vertebrate limbs are represented as torque-driven systems, including:

- Actuation in vertebrate limbs is often asymmetric—whereas engineered actuators are symmetric. That is, the angular velocities, accelerations and torques that can be produced in one rotational direction (say, flexion) are not necessarily equal to those that can be produced in the other (say, extension). While most reasonable engineers would design and build systems with symmetric actuation, most biological systems lack such symmetry. For example, compare the flexor versus extensor muscles of your fingers.
- Most muscles cross multiple DOFs. Again, most reasonable engineers would design and build systems with tendons carefully routed (often threaded through the inside of the links) to selectively actuate a single joint. But most biological limbs have *musculotendons* (the combined entity of the tendon of origin, the muscle fibers, and tendon of insertion [9]) with muscles that lie outside the bones, and tendons routed along the length of the limb actuating the multiple DOFs they cross. Thus, the individual DOFs of a vertebrate limb cannot be actuated in strict isolation.
- Most DOFs are crossed by multiple muscles. Again, while most reasonable engineers would design and build systems where only two opposing tendons drive a given DOF, the vast majority of vertebrate DOFs are not actuated in this way. Thus there is no unique set of muscle forces to produce a given net joint torque.

As we shall see, these mathematically inconvenient features of vertebrate limbs require either careful application, or extension, of the analytical approaches that were developed for torque-driven robotic systems. This situation highlights the unavoidable conceptual struggle in neuromechanics between mathematical rigor and expediency versus biological realism.

---

[4]Actuator is the generic engineering term for a motor or some other device that produces forces or mechanical work.

2.2 Forward Kinematic Analysis of Limbs

The *forward kinematics* of a limb determine the location and orientation of its *end-point* with respect to its *base*, given the relative configurations of each pair of adjacent links of the limb [6]. The base is usually the origin of the fixed, reference coordinate system—(x, y) or (x, y, z) in Fig. 2.1—chosen to represent the Cartesian coordinates of the workspace of the limb. The endpoint is the final, functional part of the limb. It is the point of interest, such as the hand when we speak of the arm for reach tasks, the foot when we speak of the legs for locomotion, or the fingertips when we speak of the hand for manipulation. Here I briefly present a simplified version of well-established methods to calculate forward kinematics of limbs. These simple kinematic formulations are common in neuromechanics studies, and sufficient to address important debates of motor control. In [6, 7] you can find an in-depth and generalized treatment of these topics.

This basic kinematic problem is: given a mathematical representation of the robotic or biological limb, and its joint angles and angular velocities, what is the position and velocity of its endpoint? To do so we must first understand how to create a mathematical representation of the forward kinematics of the limb.

Consider the example of a human arm, modeled as the planar 2 DOF serial manipulator shown in Fig. 2.1a. It is called a *planar model* because it is constrained to lie on a 2D plane; in contrast to a *spatial model* like the thumb model in Fig. 2.1b that allows motion in 3D space. The parameters of the arm model needed to calculate the position of the hand (i.e., the endpoint) are the lengths of the forearm and upper arm, and the angles of the shoulder and elbow joints as shown in Fig. 2.1a. Using the sample parameter values shown in the figure, it requires only basic knowledge of geometry to calculate the endpoint location by inspection as

$$(x, y) = (25.4 \cos(135°) + 30.5 \cos(15°), 25.4 \sin(135°) + 30.5 \sin(15°)) \quad (2.1)$$

$$(x, y) = (11.5, 25.9) \text{ in cm} \quad (2.2)$$

That was simple enough. However, now consider the 3D model of the thumb shown in Fig. 2.1b, in which there is a universal joint at both the carpometacarpal (CMC) and metacarpophalangeal (MCP) joints, and one hinge joint at the IP (interphalangeal) joint [10]. Say the metacarpal bone (closest to the wrist) has length 5.08 cm, the proximal phalanx (middle bone) has length 3.18 cm, and the distal

	Joint	Angle
Table 2.1 Sample joint angles for the spatial thumb model in Fig. 2.1b	CMC adduction-abduction	$-45°$
	CMC flexion-extension	$20°$
	MCP adduction-abduction	$-10°$
	MCP flexion-extension	$-30°$
	IP flexion-extension	$-20°$

phalanx (the bone on the thumbtip) has length 2.54 cm, and the joint angles are as in Table 2.1. Where is the endpoint then?

The mathematical expression for calculating the thumb endpoint coordinates for any set of joint angles is quite complicated, and even difficult to calculate by inspection. However, we are able to calculate the endpoint position in relation to a base frame in a systematic way if we use *homogeneous transformations*. Appendix A provides a brief introduction to these tools. It is imperative that you read it before continuing if you have not worked with the fundamentals of linear algebra or robot kinematics recently.

2.3 The Forward Kinematic Model

A limb is an *open kinematic chain* because it is a serial arrangement of articulated rigid bodies, Fig. 2.2. The posture of the limb is determined by its kinematic DOFs of the system, defined in this book by variables $q_1, q_2, q_3, \ldots, q_N$, that are also called *generalized coordinates* in mechanical analysis. In the case where anatomical joints are assumed to be rotational joints, the generalized coordinates are angles; but they can also be linear displacements for prismatic joints in robotic systems or in anatomical joints that can slide. As mentioned above, using pure rotational joints (pin, universal, or ball-and-socket joints) is common in musculoskeletal models [5], but it is a critical assumption that can have important consequences to the validity and utility of the model [2–4, 11].

Based on the techniques presented in Appendix A, a forward kinematic model of a limb is created using the following steps:

1. *Create the necessary homogenous transformation matrices, one for each DOF.* Take Fig. 2.2 as an example. Recall that we describe rigid bodies by attaching a *frame of reference* to each body, and homogeneous transformations are used to relate adjacent frames of reference. From now on, we do not speak of the rigid links any more, but only treat the frames of reference. This allows you to find[5]

$$T_{base}^{endpoint} = T_0^N \qquad (2.3)$$

---

[5] A note about typesetting conventions set forth in Appendix A. Capital letters as superscripts or subscripts (italicized or not) like M or N indicate extremes of ranges. Thus the endpoint of a limb is assigned frame N, and dimensionality of a vector or matrix are $\mathbf{v} \in \mathbb{R}^N$ or $A \in \mathbb{R}^{M \times N}$, respectively. Indices that are lowercase italicized letters like n, i, or j signify a number within a range. The letter M need not stand for muscles, or n for an intermediate frame of reference. They are simply letters to indicate dimensions and indices, and change with the context of the material. Vectors are lowercase letters typeset as $\mathbf{v}$, which can be also specified to be expressed in a given frame of reference, say frame 0, as $\mathbf{v}_0$. Or if the start and end of a vector are specified, it will be typeset as $\mathbf{p}_{0,N}$. Matrices are written as *italicized* upper case letters, such at the matrix T, which can also carry subscripts and superscripts depending on their meaning like $T_{base}^{endpoint}$. I use lowercase italics for general scalars (i.e., numbers).

If there is a single DOF between each rigid body (like pin joints in the figure), there is usually one frame of reference per link, with one homogeneous transformation per DOF—a total of $N - 1$ in this case. However, this example requires N homogeneous N transformations because the last frame of reference is needed to describe the location and orientation of the endpoint with respect to frame $N - 1$. But there are no DOFs between frames $N - 1$ and N as both frames are fixed to the same rigid body. The addition of such extra (or 'dummy') frames of reference is sometimes necessary to define the forward kinematic model of the limb. To avoid confusion, the end of a range will always be a capital letter like N. Thus,

$$T_0^N = T_0^1 \ T_1^2 \ \ldots \ T_{N-2}^{N-1} \ T_{N-1}^N \tag{2.4}$$

where

$$T_0^N = \begin{bmatrix} R_0^N & \mathbf{p}_{0,N} \\ 0 \ 0 \ 0 & 1 \end{bmatrix} \tag{2.5}$$

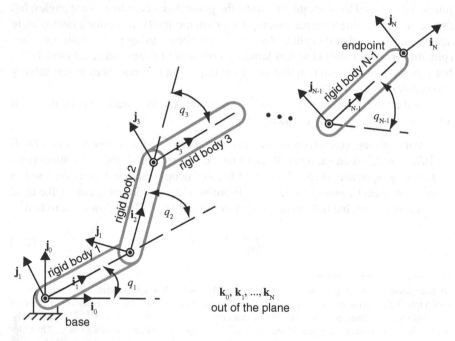

Fig. 2.2 In this book, limbs are modeled as serial linkage mechanisms (also called open kinematic chains). When defining the kinematics of such arrangements of rigid bodies, it is typical to attach a unique frame of reference to each rigid body, and thereafter perform the analysis on those frames of reference. However, this example has N frames of reference but only $N - 1$ rigid bodies and DOFs. The last frame of reference is used to describe the location and orientation of the endpoint with respect to frame $N - 1$

If there are 2 or more DOFs between two rigid bodies, like the CMC joint at the base of the thumb in Fig. 2.1, then intermediate frames of reference, and their respective homogeneous transformations, are needed to represent these DOFs. Appendix A and Sect. 2.4 discuss the importance of defining and allocating the DOFs of a limb in a specific order. See, for example, the kinematic models of the thumb in [10, 11].

2. *Extract the position of the endpoint from the homogeneous transformation* T_0^N. Note that in Eq. 2.5 the vector $\mathbf{p}_0^N$ is the location of the endpoint with respect to the base.

3. *Extract the orientation of the endpoint from the homogeneous transformation* T_0^N. The orientation of the last link is given by the matrix R_0^N. But notice that the limb in Fig. 2.2 is generic enough that, by having many DOFs, it raises the issue of kinematic redundancy. More on this in Sect. 7.1.

2.4 Application of the Forward Kinematic Model to a Simple Planar Limb

Consider the planar, 2 DOF, limb shown in Fig. 2.3. This limb is anchored to ground, where ground is the base frame, or frame 0. The first DOF (i.e., q_1) rotates the first link in a positive (as per the right-hand-rule) direction. The second DOF (i.e., q_2) *rotates the second link with respect to the first link* as shown in Fig. 2.2. This is an important convention in kinematic analysis in robotics: the generalized rotational coordinates q_i are relative to the prior body. Other branches of engineering and physics prefer all angles to be measured with respect to the base frame. But it will become apparent

Fig. 2.3 A 2-link, 2 DOF planar serial kinematic chain with its body-fixed frames of reference. Note that, even though there is no third DOF, we add the third 'dummy' frame of reference to place the frame of reference of the endpoint at the end of the second link

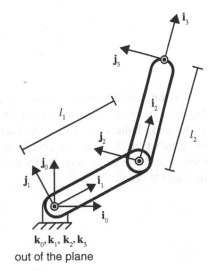

$\mathbf{k}_0, \mathbf{k}_1, \mathbf{k}_2, \mathbf{k}_3$
out of the plane

why this simplifies the arithmetic of this kinematic analysis. Importantly, this also means that we need to define a reference posture of the limb for which all angles are 0, as all adjacent frames of reference are aligned with each other. For this we use the configuration where the kinematic chain is straight and aligned with the horizontal axis of the base frame, by definition making all q_i equal to zero, all $\mathbf{i}_i$ parallel and pointing to the right, all $\mathbf{j}_i$ pointing up, and all $\mathbf{k}_i$ coming out of the page.

The first question is to define T_0^1 as described in Appendix A. Well, we see that it is a rotation about the first joint, where the origin of the first joint is the same as the origin of the base frame. This transformation has a rotation about the $\mathbf{k}_0$ axis of magnitude q_1, and no translation.

Go ahead and use Eq. A.35 to obtain the following matrices, where for succinctness I use the common shorthand of c_1 for $cos(q_1)$, s_1 for $sin(q_1)$; and c_{12}, s_{12} for cos $(q_1 + q_2)$ and $sin(q_1 + q_2)$, respectively,

$$T_0^1 = \begin{bmatrix} c_1 & -s_1 & 0 & 0 \\ s_1 & c_1 & 0 & 0 \\ 0 & 0 & 1 & 0 \\ 0 & 0 & 0 & 1 \end{bmatrix} \tag{2.6}$$

$$T_1^2 = \begin{bmatrix} c_2 & -s_2 & 0 & l_1 \\ s_2 & c_2 & 0 & 0 \\ 0 & 0 & 1 & 0 \\ 0 & 0 & 0 & 1 \end{bmatrix} \tag{2.7}$$

$$T_2^3 = \begin{bmatrix} 1 & 0 & 0 & l_2 \\ 0 & 1 & 0 & 0 \\ 0 & 0 & 1 & 0 \\ 0 & 0 & 0 & 1 \end{bmatrix} \tag{2.8}$$

It is important to point out that, even though there is no third DOF q_3, we add the third 'dummy' transformation T_2^3 to place the frame of reference of the endpoint at the end of the second link. As mentioned in Sect. 2.3, one should not confuse the number of DOFs of the system with the number of kinematic transformations used to describe the system. We then concatenate all three coordinate transformation matrices to obtain

$$T_0^3 = T_0^1 T_1^2 T_2^3 = \begin{bmatrix} c_1 c_2 - s_1 s_2 & -c_1 s_2 - s_1 c_2 & 0 & c_1(l_2 c_2 + l_1) - s_1(l_2 s_2) \\ s_1 c_2 + c_1 s_2 & -s_1 s_2 + c_1 c_2 & 0 & s_1(l_2 c_2 + l_1) + c_1(l_2 s_2) \\ 0 & 0 & 1 & 0 \\ 0 & 0 & 0 & 1 \end{bmatrix}$$

(2.9)

which with the help of the trigonometric identities

$$\cos(\alpha \pm \beta) = \cos \alpha \cos \beta \mp \sin \alpha \sin \beta \tag{2.10}$$
$$\sin(\alpha \pm \beta) = \sin \alpha \cos \beta \pm \cos \alpha \sin \beta \tag{2.11}$$

simplifies to

$$T_0^3 = T_0^1 T_1^2 T_2^3 = \begin{bmatrix} c_{12} & -s_{12} & 0 & l_1 c_1 + l_2 c_{12} \\ s_{12} & c_{12} & 0 & l_1 s_1 + l_2 s_{12} \\ 0 & 0 & 1 & 0 \\ 0 & 0 & 0 & 1 \end{bmatrix}$$

(2.12)

It follows that

$$R_0^3 = \begin{bmatrix} c_{12} & -s_{12} & 0 \\ s_{12} & c_{12} & 0 \\ 0 & 0 & 1 \end{bmatrix}$$

(2.13)

and

$$\mathbf{p}_{0,3} = \begin{pmatrix} l_1 c_1 + l_2 c_{12} \\ l_1 s_1 + l_2 s_{12} \\ 0 \end{pmatrix}$$

(2.14)

Look closely at Eq. 2.13 and notice that its structure says that it represents a right-handed rotation about the $\mathbf{k}_0$ axis of a magnitude equal to $q_1 + q_2$. Also, there are no other rotations, as is expected from a planar limb. Similarly, the vector $\mathbf{p}_{0,3}$ represents a displacement on the $\mathbf{i}_0$—$\mathbf{j}_0$ plane, with no component in the $\mathbf{k}_0$ direction, also as expected for a planar limb.

Therefore, in this case the forward kinematic model (also called the geometric model), $G(\mathbf{q})$, is

Fig. 2.4 The forward
kinematic model of the
planar limb in Fig. 2.3 in
more familiar Cartesian
coordinates

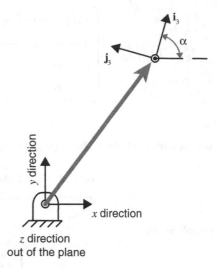

$$G(\mathbf{q}) = \begin{pmatrix} \text{displacement in } \mathbf{i}_0 \text{ direction} \\ \text{displacement in } \mathbf{j}_0 \text{ direction} \\ \text{rotation about the } \mathbf{k}_0 \text{ axis} \end{pmatrix} = \begin{pmatrix} l_1 c_1 + l_2 c_{12} \\ l_1 s_1 + l_2 s_{12} \\ q_1 + q_2 \end{pmatrix} \qquad (2.15)$$

This example raises an important concept in retrospect: How many kinematic DOFs does a rigid body have on the plane? The answer is three, which are two displacements and one rotation—as revealed by the elements of T_0^3 and written out explicitly in $G(\mathbf{q})$. We will explore the relationships between the kinematic DOFs of the endpoint and the kinematic DOFs of the limb later in this book. But for now, this simple example allows us to validate this intuition because you could have just as easily written out the forward kinematic model of this limb by inspection. Try it.

But the definition of the forward kinematic model in Eq. 2.15 holds other surprises. On one hand, creating this vector function is natural as it describes the position and orientation of the endpoint frame of reference. On the other hand, you may have probably never seen a vector with mixed units. The first two elements are distances, and the last element is an angle.

Such vectors are necessary to express forward kinematic models—they are a natural consequence of the homogeneous transformation matrices that contain a combination of rotations and displacements [6]. As you will soon see this gives rise to the very convenient mathematical formulations of twists and wrenches presented in Sect. 2.6.

For the sake of clarity, I define that forward kinematic vector as $\mathbf{x}$ using the more familiar variables shown in Fig. 2.4 as follows

$$\mathbf{x} = \begin{pmatrix} x \\ y \\ \alpha \end{pmatrix} = G(\mathbf{q}) = \begin{pmatrix} G_x(\mathbf{q}) \\ G_y(\mathbf{q}) \\ G_\alpha(\mathbf{q}) \end{pmatrix} = \begin{pmatrix} l_1 c_1 + l_2 c_{12} \\ l_1 s_1 + l_2 s_{12} \\ q_1 + q_2 \end{pmatrix} \qquad (2.16)$$

2.5 Using the Forward Kinematic Model to Obtain Endpoint Velocities

How is the velocity of the endpoint ($\dot{x}$) related to the angular velocities of the joints ($\dot{q}$)? Note that a dot above a variable is a shorthand that indicates the time derivative, thus $\dot{a} = \frac{da}{dt}$. This is also part of the forward kinematics problem because the joint angular velocities are inputs that produce the endpoint velocities as an output.

Consider the case of a limb with N kinematic DOFs, where we know that the forward kinematic model specifies the location and orientation of the endpoint as a function of the joint angles

$$\mathbf{x} = G(\mathbf{q}) \tag{2.17}$$

We now want to know the time derivative of the forward kinematic model such that

$$\dot{\mathbf{x}} = \frac{G(\mathbf{q})}{dt} = \frac{\partial G(\mathbf{q})}{\partial \mathbf{q}} \frac{d\mathbf{q}}{dt} = \frac{\partial G(\mathbf{q})}{\partial \mathbf{q}} \dot{\mathbf{q}} \tag{2.18}$$

For a specific example where

$$\mathbf{x} = \begin{pmatrix} x \\ y \\ \alpha \end{pmatrix} \tag{2.19}$$

$$\mathbf{q} = \begin{pmatrix} q_1 \\ q_2 \\ \vdots \\ q_N \end{pmatrix} \tag{2.20}$$

$$\dot{\mathbf{x}} = \begin{pmatrix} \dot{x} \\ \dot{y} \\ \dot{\alpha} \end{pmatrix} \tag{2.21}$$

$$\dot{\mathbf{q}} = \begin{pmatrix} \dot{q}_1 \\ \dot{q}_2 \\ \vdots \\ \dot{q}_N \end{pmatrix} \tag{2.22}$$

The definition of the partial derivatives of the vector function $G(\mathbf{q})$ is

$$\frac{\partial G(\mathbf{q})}{\partial \mathbf{q}} = J(\mathbf{q}) = \begin{bmatrix} \frac{\partial G_x(\mathbf{q})}{\partial q_1} & \frac{\partial G_x(\mathbf{q})}{\partial q_2} & \cdots & \frac{\partial G_x(\mathbf{q})}{\partial q_N} \\ \frac{\partial G_y(\mathbf{q})}{\partial q_1} & \frac{\partial G_y(\mathbf{q})}{\partial q_2} & \cdots & \frac{\partial G_y(\mathbf{q})}{\partial q_N} \\ \frac{\partial G_\alpha(\mathbf{q})}{\partial q_1} & \frac{\partial G_\alpha(\mathbf{q})}{\partial q_2} & \cdots & \frac{\partial G_\alpha(\mathbf{q})}{\partial q_N} \end{bmatrix} \qquad (2.23)$$

where N is the number of DOFs, and $J(\mathbf{q})$ is called the *Jacobian* of the system. In this case $J \in \mathbb{R}^{3 \times N}$. The instantaneous 3D endpoint velocity vector can be calculated using the following equation:

$$\dot{\mathbf{x}} = J(\mathbf{q})\,\dot{\mathbf{q}} \qquad (2.24)$$

And when the Jacobian is invertible (see Sect. 3.5) we can find the instantaneous joint angular velocities associated with a given endpoint velocity vector using the following equation

$$\dot{\mathbf{q}} = J(\mathbf{q})^{-1}\,\dot{\mathbf{x}} \qquad (2.25)$$

2.6 General Case of the Jacobian in the Context of Screws, Twists, and Wrenches

The Jacobian of a serial linkage system is fundamental to the calculation of the feasible motions and forces that it can produce. The general definition of a Jacobian needs to address the fact that the endpoint of the kinematic chain, as a rigid body, has 6 DOFs: three translations and three rotations. In the formal kinematics of rigid body mechanics [12], this falls within the field of screw theory [6]. Such a combined vector is called a *screw*. It consists of a pair of 3D vectors, in this case the translations and rotations a rigid body can have—which I still call $\mathbf{x}$ because it represents the forward kinematic model of the endpoint as in Eq. 2.16.

In the general case where we consider all 6 DOFs of the frame of reference fixed at the endpoint,

$$\mathbf{x} = \begin{pmatrix} \begin{pmatrix} x \\ y \\ z \end{pmatrix} \\ \begin{pmatrix} \alpha \\ \beta \\ \gamma \end{pmatrix} \end{pmatrix} \in \mathbb{R}^6 \qquad (2.26)$$

containing the 3D position and orientation vectors.

The time derivative of this positional/rotational screw vector is the *twist* vector of linear and angular velocities, respectively—which I call $\dot{\mathbf{x}}$

$$\dot{\mathbf{x}} = \left(\left(\begin{array}{c} \dot{x} \\ \dot{y} \\ \dot{z} \\ \dot{\alpha} \\ \dot{\beta} \\ \dot{\gamma} \end{array} \right) \right) \tag{2.27}$$

As we will see further on, this screw concept that combines elements of different units extends to the force and torque vectors an endpoint can produce—called the endpoint *wrench* vector [13]

$$\mathbf{w} = \left(\left(\begin{array}{c} f_x \\ f_y \\ f_z \\ \tau_\alpha \\ \tau_\beta \\ \tau_\gamma \end{array} \right) \right) \tag{2.28}$$

where f and τ are the components of force and torque along their respective dimensions. This means that in the general case the full Jacobian has 6 rows and N columns [14], $J \in \mathbb{R}^{6 \times N}$

$$\frac{\partial G(\mathbf{q})}{\partial \mathbf{q}} = J(\mathbf{q}) = \begin{bmatrix} \frac{\partial G_x(\mathbf{q})}{\partial q_1} & \frac{\partial G_x(\mathbf{q})}{\partial q_2} & \cdots & \frac{\partial G_x(\mathbf{q})}{\partial q_N} \\ \frac{\partial G_y(\mathbf{q})}{\partial q_1} & \frac{\partial G_y(\mathbf{q})}{\partial q_2} & \cdots & \frac{\partial G_y(\mathbf{q})}{\partial q_N} \\ \frac{\partial G_z(\mathbf{q})}{\partial q_1} & \frac{\partial G_z(\mathbf{q})}{\partial q_2} & \cdots & \frac{\partial G_z(\mathbf{q})}{\partial q_N} \\ \frac{\partial G_\alpha(\mathbf{q})}{\partial q_1} & \frac{\partial G_\alpha(\mathbf{q})}{\partial q_2} & \cdots & \frac{\partial G_\alpha(\mathbf{q})}{\partial q_N} \\ \frac{\partial G_\beta(\mathbf{q})}{\partial q_1} & \frac{\partial G_\beta(\mathbf{q})}{\partial q_2} & \cdots & \frac{\partial G_\beta(\mathbf{q})}{\partial q_N} \\ \frac{\partial G_\gamma(\mathbf{q})}{\partial q_1} & \frac{\partial G_\gamma(\mathbf{q})}{\partial q_2} & \cdots & \frac{\partial G_\gamma(\mathbf{q})}{\partial q_N} \end{bmatrix} \tag{2.29}$$

This succinct presentation of the general case of the Jacobian matrix for a limb raises several questions, some beyond the scope of this book. For example:

- How does one find the 6D screw vector for a generic robotic or biological limb? In [6, 7] you can find examples of this, but their derivation requires a working knowledge of kinematics.
- What is the relationship between the N kinematic DOFs of the limb and the 6 DOFs the end link can have? As you shall see in later chapters, roboticists are very mindful to design limbs with 6 or fewer DOFs so that the Jacobian matrix is easier

to compute and manipulate. But there are others who design, say, snake robots, that have many more kinematic DOFs (see discussion of kinematic redundancy in Sect. 7.1). Similarly, biological limbs are often analyzed as having 6 or fewer kinematic DOFs [5].

But as mentioned above, the types of simplified limbs presented in this book are common in neuromechanics studies, and suffice to address important debates of motor control. My goal is to present simplified systems to build intuition that can be carried forward to more complex (i.e., anatomically realistic) limbs.

2.7 Using the Jacobian of a Planar System to Find Endpoint Velocities

Many of the examples and cases we investigate will use planar systems without the loss of generality. This will make the presentation of concepts easier to describe and illustrate. In the case of a planar 2-link, 2-joint system as in Fig. 2.3, the 2D forward kinematic model for the endpoint is

$$\mathbf{x} = \begin{pmatrix} x \\ y \\ \alpha \end{pmatrix} = \begin{pmatrix} G_x(\mathbf{q}) \\ G_y(\mathbf{q}) \\ G_\alpha(\mathbf{q}) \end{pmatrix} = \begin{pmatrix} l_1 c_1 + l_2 c_{12} \\ l_1 s_1 + l_2 s_{12} \\ q_1 + q_2 \end{pmatrix} \qquad (2.30)$$

Taking the appropriate partial derivates produces

$$J(\mathbf{q}) = \begin{bmatrix} -l_1 s_1 - l_2 s_{12} & -l_2 s_{12} \\ l_1 c_1 + l_2 c_{12} & l_2 c_{12} \\ 1 & 1 \end{bmatrix} \qquad (2.31)$$

where

$$\dot{\mathbf{x}} = \begin{pmatrix} \dot{x} \\ \dot{y} \\ \dot{\alpha} \end{pmatrix} = \begin{bmatrix} -l_1 s_1 - l_2 s_{12} & -l_2 s_{12} \\ l_1 c_1 + l_2 c_{12} & l_2 c_{12} \\ 1 & 1 \end{bmatrix} \begin{pmatrix} \dot{q}_1 \\ \dot{q}_2 \end{pmatrix} \qquad (2.32)$$

As shown in Eq. 2.32, and graphically in Fig. 2.5, each column of the Jacobian is the instantaneous endpoint velocity vector produced by one unit of the corresponding joint angular velocity (i.e., the first column of the Jacobian is the endpoint velocity vector produced by an angular velocity of 1 rad/s at the first joint if other joint angular velocities are zero, the second column is the endpoint velocity vector produced by a 1 rad/s angular velocity at the second joint if other joint angular velocities are zero, etc.). If there are simultaneous angular velocities at both joints, their *instantaneous* effects at the endpoint simply add linearly. The last row says that the angular velocity of the endpoint is simply the sum of the angular velocity at each joint.

$$J\,(135°,\ -120°) = \begin{bmatrix} -0.2590 & -0.0787 \\ 0.1550 & 0.2950 \\ 1 & 1 \end{bmatrix}$$ Endpoint velocity direction produced by each joint's positive angular velocity

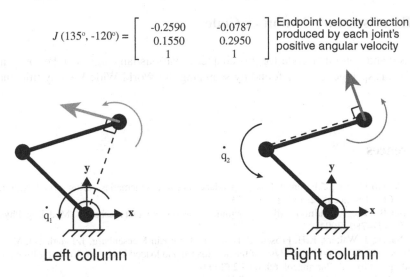

Left column   Right column

Fig. 2.5 Illustration of the Jacobian for a 2 DOF planar limb. For the posture shown, the columns of the 2×2 Jacobian show the expected *instantaneous* endpoint linear and angular velocity for isolated angular velocities of 1 rad/s at each of the joints. If both joints are actuated, then their contribution to instantaneous endpoint velocity simply add. The limb parameters are, as per the convention in Fig. 2.3, $l_i = 25.4$ cm, $l_2 = 30.5$ cm, $q_1 = 135°$, and $q_2 = -120°$

But let us look at the planar arm example in Fig. 2.5 and Eq. 2.32 in detail. First, we notice that (other than the last row) the values of the elements of the Jacobian matrix can be posture dependent (i.e., they change as the posture—or angles q_1 and q_2—change). Second, we are therefore forced to always speak of *instantaneous* endpoint velocities because these values only hold for that posture, and the posture is changing—by definition—given that the joints have angular velocities. And third, given that the forward kinematic model and the Jacobian involve trigonometric functions, the mapping from angular velocities to endpoint velocities changes in nonlinear ways as the motion progresses. This will naturally make the dynamical control of such systems complex because the properties of the system change in nonlinear ways as the system moves. However, for a given posture (as in the case of static force production described later), both the forward kinematic model and the Jacobian are fixed.

Further treatment of the Jacobian can be found in [6, 14]. For now, it suffices for the reader to know that the Jacobian relates joint velocities to endpoint velocities, and that it can be derived in a straightforward manner for any arbitrary serial manipulator by taking partial derivatives of the analytical expressions for the forward kinematic model. These can be derived from first principles either using homogeneous coordinate transformations, or by using other methods such as D-H parameterization, among others.

2.8 Exercises and Computer Code

Exercises and computer code for this chapter in various languages can be found at http://extras.springer.com or found by searching the World Wide Web by title and author.

References

1. F.J. Valero-Cuevas, C.F. Small, Load dependence in carpal kinematics during wrist flexion in vivo. Clin. Biomech. **12**, 154–159 (1997)
2. Y. Bei, B.J. Fregly, Multibody dynamic simulation of knee contact mechanics. Med. Eng. Phys. **26**(9), 777–789 (2004)
3. H. Nagerl, J. Walters, K.H. Frosch, C. Dumont, D. Kubein-Meesenburg, J. Fanghanel, M.M. Wachowski, Knee motion analysis of the non-loaded and loaded knee: a re-look at rolling and sliding. J. Physiol. Pharmacol. **60**, 69–72 (2009)
4. C.E. Wall, A model of temporomandibular joint function in anthropoid primates based on condylar movements during mastication. Am. J. Phys. Anthropol. **109**(1), 67–88 (1999)
5. F.J. Valero-Cuevas, H. Hoffmann, M.U. Kurse, J.J. Kutch, E.A. Theodorou, Computational models for neuromuscular function. IEEE Rev. Biomed. Eng. **2**, 110–135 (2009)
6. R.M. Murray, Z. Li, S.S. Sastry, *A Mathematical Introduction to Robotic Manipulation* (CRC, Boca Raton, 1994)
7. T. Yoshikawa, *Foundations of Robotics: Analysis and Control* (MIT Press, Cambridge, 1990)
8. F.J. Valero-Cuevas, A mathematical approach to the mechanical capabilities of limbs and fingers. Adv. Exp. Med. Biol. **629**, 619–633 (2009)
9. F.E. Zajac, Muscle and tendon: properties, models, scaling, and application to biomechanics and motor control. Crit. Rev. Biomed. Eng. **17**(4), 359–411 (1989)
10. V.J. Santos, F.J. Valero-Cuevas, Reported anatomical variability naturally leads to multimodal distributions of Denavit-Hartenberg parameters for the human thumb. IEEE Trans. Biomed. Eng. **53**, 155–163 (2006)
11. F.J. Valero-Cuevas, M.E. Johanson, J.D. Towles, Towards a realistic biomechanical model of the thumb: the choice of kinematic description may be more critical than the solution method or the variability/uncertainty of musculoskeletal parameters. J. Biomech. **36**, 1019–1030 (2003)
12. O. Bottema, B. Roth, *Theoretical Kinematics* (Dover Publications, New York, 2012)
13. Wikipedia contributors, Basis vectors, Wikipedia, The Free Encyclopedia. https://en.wikipedia.org/wiki/Screw_theory. Accessed 12 Feb 2015
14. T. Yoshikawa, Translational and rotational manipulability of robotic manipulators, in *1991 International Conference on Industrial Electronics, Control and Instrumentation* (IEEE, 2002), pp. 1170–1175

Chapter 3
Limb Mechanics

Abstract Limb mechanics involve limb kinematics, and the forces and torques that cause limb loading and motion. Mechanics can be both static and dynamic depending on whether motion is prevented or not, respectively. Studying limb motions that result from applied forces and torques falls within the realm of rigid-body dynamics, which is a specialized branch of mechanics. However, I will mostly consider the case of static mechanics because it suffices to illustrate and debate important concepts in neuromechanics. This chapter focuses on presenting some fundamental concepts of how limbs produce static forces.

3.1 Derivation of the Relationship Between Static Endpoint Forces and Joint Torques

Static mechanics is accessible to readers with a variety of scientific and non-scientific backgrounds because it is based on geometry. This makes it possible for the reader to apply the concepts presented in this book to many problems of interest in neuromechanics. Other excellent texts present in detail the areas of static mechanics of robotic limbs [1], and the problem of rigid-body dynamics in the context of robotic [1–3] and neuromuscular limbs [4].

Defining the static force production capabilities of limbs begins by deriving the equation that determines the production of wrenches at the endpoint as a function of joint torques. Recall that the mechanical output of the last link of a limb can be both forces and torques (i.e., the endpoint wrench from Eq. 2.28) after [1]. This equation is derived using the principle of virtual work, a corollary of the law of conservation of energy.

We begin by stating the law of conservation of energy between the internal and external work that the system produces. For simplicity, I present this for a pure force at the endpoint and for torques at the joints. Recall that mechanical work is the product of force times displacement. In the linear and rotational cases, respectively, this is expressed as follows

© Springer-Verlag London 2016
F.J. Valero-Cuevas, *Fundamentals of Neuromechanics*,
Biosystems & Biorobotics 8, DOI 10.1007/978-1-4471-6747-1_3

Fig. 3.1 Geometric
relationship between the
displacement of the
endpoint, $\Delta\mathbf{x}$, and the
rotation of the joints, $\Delta\mathbf{q}$

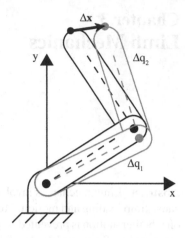

$$\text{External work} = \mathbf{f} \cdot \Delta\mathbf{x} \tag{3.1}$$
$$\text{Internal work} = \boldsymbol{\tau} \cdot \Delta\mathbf{q} \tag{3.2}$$

The external work is the force vector at the endpoint $\mathbf{f}$ times the infinitesimal displacement at the endpoint $\Delta\mathbf{x} = (\Delta x, \Delta y)^T$. The internal work is the joint torque vector $\boldsymbol{\tau}$ times the associated infinitesimal rotation at the joints $\Delta\mathbf{q} = (\Delta q_1, \Delta q_2)^T$. Figure 3.1 shows the geometric relationship between the endpoint displacement and the associated joint rotations.

Before proceeding further, we must explain several caveats. This derivation can be extended to a wrench and a screw displacement at the endpoint. The use of the dot product ensures that only forces compatible with possible displacements are considered. Similarly, the limb may have linear DOFs (like sliders) so the internal work can be from forces and displacements instead of only torques and rotations. Last, to be precise and compatible with Castigliano's theorems of virtual work, we need to speak of infinitesimal displacements (hence the term *virtual work*) so that the shape of the system (i.e., posture of the limb) does not change during the analysis.

An expression of the principle of conservation of energy is that the external and internal work must be equal in the absence of deformation of the rigid links, leading to

$$\mathbf{f} \cdot \Delta\mathbf{x} = \boldsymbol{\tau} \cdot \Delta\mathbf{q} \tag{3.3}$$

Given that we are considering infinitesimal displacements, we can take the differentials on both sides to get

$$\mathbf{f} \cdot \delta\mathbf{x} = \boldsymbol{\tau} \cdot \delta\mathbf{q} \tag{3.4}$$

Substituting the dot product operator by its equivalent vector inner product we get

$$\mathbf{f}^T \delta\mathbf{x} = \boldsymbol{\tau}^T \delta\mathbf{q} \tag{3.5}$$

If we normalize by time by dividing both sides by δt we get

$$\mathbf{f}^T \dot{\mathbf{x}} = \boldsymbol{\tau}^T \dot{\mathbf{q}} \tag{3.6}$$

Now, since we have already seen in Eq. 2.24 that $\dot{\mathbf{x}} = J(\mathbf{q})\,\dot{\mathbf{q}}$, we substitute and obtain

$$\mathbf{f}^T J(\mathbf{q})\,\dot{\mathbf{q}} = \boldsymbol{\tau}^T \dot{\mathbf{q}} \tag{3.7}$$

$$\mathbf{f}^T J(\mathbf{q}) = \boldsymbol{\tau}^T \tag{3.8}$$

And following the rules of transposition for matrix multiplication we get the main result

$$\boldsymbol{\tau} = J(\mathbf{q})^T \mathbf{f} \tag{3.9}$$

or

$$\mathbf{f} = J(\mathbf{q})^{-T} \boldsymbol{\tau} \tag{3.10}$$

more formally, these equations can be written to consider an endpoint wrench $\mathbf{w}$

$$\boldsymbol{\tau} = J(\mathbf{q})^T \mathbf{w} \tag{3.11}$$

or

$$\mathbf{w} = J(\mathbf{q})^{-T} \boldsymbol{\tau} \tag{3.12}$$

3.2 Symbolic Example Finding All Permutations of J for a Planar 2 DOF Limb

You now see that the Jacobian J of the limb is used in a variety of forms: transposed (J^T), inverted (J^{-1}), and the inverse of the transposed (J^{-T}). It is important to first get a sense of what each of these forms of the Jacobian convey. This raises a question that is addressed in the remainder of this chapter: When is the Jacobian invertible, and why? But for now, I will consider a truncated version of the Jacobian in Sect. 2.7 that is both square and invertible. In the planar limb in Fig. 2.3 we can remove α, the orientation of the last limb, as an endpoint DOF and define the forward kinematic model to only consider the displacements of the endpoint

$$\mathbf{x} = \begin{pmatrix} x \\ y \end{pmatrix} = \begin{pmatrix} G_x(\mathbf{q}) \\ G_y(\mathbf{q}) \end{pmatrix} = \begin{pmatrix} l_1 c_1 + l_2 c_{12} \\ l_1 s_1 + l_2 s_{12} \end{pmatrix} \tag{3.13}$$

Taking the appropriate partial derivates produces

$$J(\mathbf{q}) = \begin{bmatrix} -l_1 s_1 - l_2 s_{12} & -l_2 s_{12} \\ l_1 c_1 + l_2 c_{12} & l_2 c_{12} \end{bmatrix} \tag{3.14}$$

In this case, $J \in \mathbb{R}^{2 \times 2}$, and is a full rank invertible matrix (more on this in Sect. 3.5). Then, either by hand or using computer programs, you can find the following symbolic equations:

$$G(\mathbf{q}) = \begin{bmatrix} l_2 \cos(q_1 + q_2) + l_1 \cos(q_1) \\ l_2 \sin(q_1 + q_2) + l_1 \sin(q_1) \end{bmatrix}$$

$$J(\mathbf{q}) = \begin{bmatrix} -l_2 \sin(q_1 + q_2) - l_1 \sin(q_1) & -l_2 \sin(q_1 + q_2) \\ l_2 \cos(q_1 + q_2) + l_1 \cos(q_1) & l_2 \cos(q_1 + q_2) \end{bmatrix}$$

$$J^T(\mathbf{q}) = \begin{bmatrix} -\sin(q_1) l_1 - \sin(q_1 + q_2) l_2 & \cos(q_1) l_1 + \cos(q_1 + q_2) l_2 \\ -\sin(q_1 + q_2) l_2 & \cos(q_1 + q_2) l_2 \end{bmatrix}$$

$$J^{-1}(\mathbf{q}) = \begin{bmatrix} -\dfrac{\cos(q_1 + q_2)}{l_1 \cos(q_1 + q_2) \sin(q_1) - l_1 \sin(q_1 + q_2) \cos(q_1)} & -\dfrac{\sin(q_1 + q_2)}{l_1 \cos(q_1 + q_2) \sin(q_1) - l_1 \sin(q_1 + q_2) \cos(q_1)} \\ \dfrac{l_2 \cos(q_1 + q_2) + l_1 \cos(q_1)}{l_1 l_2 \cos(q_1 + q_2) \sin(q_1) - l_1 l_2 \sin(q_1 + q_2) \cos(q_1)} & \dfrac{l_2 \sin(q_1 + q_2) + l_1 \sin(q_1)}{l_1 l_2 \cos(q_1 + q_2) \sin(q_1) - l_1 l_2 \sin(q_1 + q_2) \cos(q_1)} \end{bmatrix}$$

$$J^{-T}(\mathbf{q}) = \begin{bmatrix} \dfrac{\cos(q_1 + q_2)}{\cos(q_1) \sin(q_1 + q_2) l_1 - \sin(q_1) \cos(q_1 + q_2) l_1} & -\dfrac{\cos(q_1) l_1 + \cos(q_1 + q_2) l_2}{\cos(q_1) \sin(q_1 + q_2) l_1 l_2 - \sin(q_1) \cos(q_1 + q_2) l_1 l_2} \\ \dfrac{\sin(q_1 + q_2)}{\cos(q_1) \sin(q_1 + q_2) l_1 - \sin(q_1) \cos(q_1 + q_2) l_1} & -\dfrac{\sin(q_1) l_1 + \sin(q_1 + q_2) l_2}{\cos(q_1) \sin(q_1 + q_2) l_1 l_2 - \sin(q_1) \cos(q_1 + q_2) l_1 l_2} \end{bmatrix}$$

3.3 Numerical Example Finding All Permutations of J for a Planar 2 DOF Limb

One again, using the MATLAB code, we assign values of unit link lengths and the second joint flexed 90° (in radians)

$$l_1 = 1.0$$
$$l_2 = 1.0$$
$$q_1 = 0.0$$
$$q_2 = 1.5708$$

which produces the following numerical results:

$$G = \begin{pmatrix} 1.0 \\ 1.0 \end{pmatrix}$$

$$J = \begin{bmatrix} -1.0 & -1.0 \\ 1.0 & 0.0 \end{bmatrix}$$

$$J^T = \begin{bmatrix} -1.0 & 1.0 \\ -1.0 & 0.0 \end{bmatrix}$$

$$J^{-1} = \begin{bmatrix} 0.0 & 1.0 \\ -1.0 & -1.0 \end{bmatrix}$$

$$J^{-T} = \begin{bmatrix} 0.0 & -1.0 \\ 1.0 & -1.0 \end{bmatrix}$$

Draw such a limb and let us calculate some examples, such as applying a positive angular velocity at q_1 to find the resulting instantaneous endpoint velocity vector $\dot{x}$ using Eq. 2.24

$$\text{given } \dot{q} = \begin{pmatrix} 1.0 \\ 0.0 \end{pmatrix}$$

$$\text{produces } \dot{x} = \begin{pmatrix} -1.0 \\ 1.0 \end{pmatrix}$$

Or applying that same endpoint velocity vector $\dot{x}$ to find the resulting instantaneous angular velocity vector $\dot{q}$ using Eq. 2.25

$$\text{produces } \dot{q} = \begin{pmatrix} 1.0 \\ 0.0 \end{pmatrix}$$

Or finding what input joint torque vector τ is needed to produce a horizontal static endpoint force vector f using Eq. 3.9

$$\text{given } f = \begin{pmatrix} 1.0 \\ 0.0 \end{pmatrix}$$

$$\text{produces } \tau = \begin{pmatrix} -1.0 \\ -1.0 \end{pmatrix}$$

And applying those joint torques τ to find the resulting endpoint force f in equilibrium using Eq. 3.10

$$\text{produces } f = \begin{pmatrix} 1.0 \\ 0.0 \end{pmatrix}$$

3.4 Relationship Between J^T and the Equations of Static Equilibrium

Let us look at the planar 2 DOF arm with joint angles and link lengths such that the geometry of the system is as shown in Fig. 3.2. We can use Newton's first law stating that the sum of forces and torques equals zero to find joint torques needed to exert a static 1 N force in the negative y-direction. The simplified scalar version of this equation is $\tau_i = d_i \, ||\mathbf{f}||$, where $||\mathbf{f}||$ is the magnitude of the force applied, and d_i is the perpendicular distance from the line of action of the force to the center of rotation of the joint. Using this equation, we see that $\tau_1 = d_1 \, ||\mathbf{f}|| = -0.115$ Nm and $\tau_2 = d_2 \, ||\mathbf{f}|| = -0.295$ Nm.

However, we can also do this by simple matrix multiplication using the Jacobian seen in Fig. 2.5. After selecting the first two rows as in Sect. 3.3, and using Eq. 3.9 we obtain,

$$\boldsymbol{\tau} = J^T \mathbf{f} = \begin{bmatrix} -0.259 & 0.115 \\ -0.0787 & 0.295 \end{bmatrix} \begin{pmatrix} 0.0 \\ -1.0 \end{pmatrix} = \begin{pmatrix} -0.115 \\ -0.295 \end{pmatrix}$$

Try this same comparison using the example in Sect. 3.3, or any other of your choosing. What is clear is that J^T encapsulates Newton's first law for static equilibrium. Impressive, right? Later on you will also see that the various forms of the Jacobian encapsulate many inherent kinematic and mechanical properties of the limb.

Fig. 3.2 Calculating joint torques for a planar 2 DOF arm using the Newtonian static equilibrium approach. Force units are in N. The link lengths and posture are those in Fig. 2.5

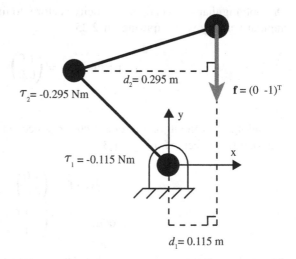

3.5 Importance of Understanding the Kinematic Degrees of Freedom of a Limb

The critical issue here is to clearly define the *independent internal versus external kinematic DOFs* of the system. The internal DOFs are the vector $\mathbf{q} \in \mathbb{R}^N$, and the external DOFs are the location and orientation of the last link (see Sect. 2.3). These concepts are closely tied to the generalized coordinates used in Lagrangian mechanics, which are the DOFs that are compatible with the kinematic constraints of the limb such that the system can do mechanical work [4–6]. Therefore, the actuation of the internal DOFs defined as $\boldsymbol{\tau} \in \mathbb{R}^N$ in Sect. 3.1 can produce work at the endpoint via twists or wrenches along the external DOFs of the endpoint as defined in Sects. 2.6 and 3.1.

Thus, as per Eq. 3.3, you should think of this as an energy input-output system where the work done by the internal DOFs maps onto the work done by the external DOFs, and vice versa. This is related to the concept of controllability—how your control signals at the internal DOFs, e.g., τ_i, affect the linear and angular coordinates, velocities, accelerations, and forces at the endpoint. In the case of torque-driven limbs, it stands to reason (although I will not prove it here as others address this in detail [1, 2]) that you can control as many external independent kinematic DOFs as you have internal independent DOFs. These external DOFs could naturally be x and y as in Eq. 2.30.

However, if you inspect Fig. 2.3 carefully, you need to see which of the DOFs of the last frame of reference (i.e., the last rigid body on the chain, which contains the endpoint of the system) are actually independent. A rigid body in the plane has *three* degrees of freedom: x and y locations, and an orientation, call it α, as we say in Eq. 2.30. But that leads to a non-square (and non-invertible) Jacobian because we only have 2 internal DOFs, q_1 and q_2.

Consider then an alternative, as shown in Eq. 3.13, where we have explicitly phrased the problem as the 2 internal DOFs controlling *only* the x and y location of the last rigid body independently of each other. Just as explicitly, its orientation α is not controlled independently, but simply follows from what x and y are. That is, α is a kinematic parameter of the last link, but not an independent DOF of it.

We could have just as easily written the forward kinematic model as

$$\mathbf{x} = \begin{pmatrix} x \\ \alpha \end{pmatrix} = \begin{pmatrix} G_x(\mathbf{q}) \\ G_\alpha(\mathbf{q}) \end{pmatrix} = \begin{pmatrix} l_1 c_1 + l_2 c_{12} \\ q_1 + q_2 \end{pmatrix} \tag{3.15}$$

where $G_\alpha(\mathbf{q})$ is the total rotation of the last link. This geometric model[1] says that we have phrased the problem as one of independently controlling the x location and orientation α of the last rigid body in the kinematic chain (remember, you can only control two). But, the remaining third kinematic parameter of the last rigid link, y, cannot be controlled independently and is no longer an independent DOF. It simply follows from what x and α are. Explain to yourself why Eq. 3.15 is as valid as the

---

[1]Note I use forward kinematic model and geometric model interchangeably.

alternative in Eq. 3.13, and what differences there are between their definitions and uses.

Which forward kinematic model is correct for this limb, Eqs. 2.30, 3.13 or 3.15? Well, all of them, depending on how you intend to use them. You now see that the definition of the forward kinematic model is a matter of preference that should be aligned with the goal of your analysis. The key fact is that if you have N internal kinematic DOFs, you are able to independently control at most N external kinematic DOFs if $N \le 6$. Why at most 6? Well, because a rigid body in space has 6 DOFs, as stated in Eq. 2.26. And how you phrase your forward kinematic model depends on which DOFs you want to define as independently controllable. In the case where you have more than 6 internal DOFS (think of a snake robot with many links), you can still control the 6 kinematic DOFs of the last link, but your system is kinematically redundant in that there is an infinite number of possible configurations $\mathbf{q}$ that can produce the same endpoint position and orientation (and you can do internal work that does not produce external work, in violation of the derivation in Sect. 3.1). We will discuss many versions of redundancy at length in later chapters. But, for the sake of expediency, in this chapter we will focus on limbs with at most 6 internal DOFs.

3.6 Analysis of a Planar 3 DOF Limb

Note that the last link of a planar limb has 3 external kinematic DOFs, as mentioned above. So let us move beyond the simpler systems shown in Fig. 2.3 and consider the system in Fig. 3.3. In this case, we have enough internal DOFs to control the 3 external DOFs at the last link: its location and orientation. The forward kinematic model is then

$$\mathbf{x} = \begin{pmatrix} x \\ y \\ \alpha \end{pmatrix} = \begin{pmatrix} G_x(\mathbf{q}) \\ G_y(\mathbf{q}) \\ G_\alpha(\mathbf{q}) \end{pmatrix} = \begin{pmatrix} l_1c_1 + l_2c_{12} + l_3c_{123} \\ l_1s_1 + l_2s_{12} + l_3s_{123} \\ q_1 + q_2 + q_3 \end{pmatrix} \tag{3.16}$$

and its Jacobian is

$$J = \begin{bmatrix} -l_1s_1 - l_2s_{12} - l_3s_{123} & -l_2s_{12} - l_3s_{123} & -l_3s_{123} \\ l_1c_1 + l_2c_{12} + l_3c_{123} & l_2c_{12} + l_3c_{123} & l_3c_{123} \\ 1 & 1 & 1 \end{bmatrix} \tag{3.17}$$

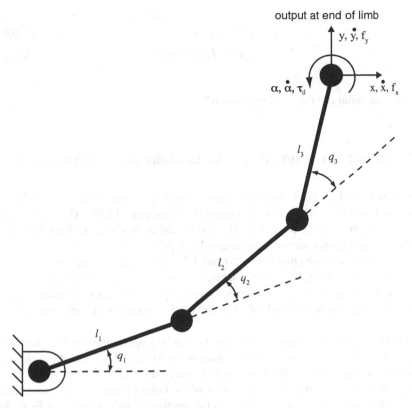

Fig. 3.3 A 3-link, 3 DOF planar serial kinematic chain. The details of the individual frames of reference are not shown for clarity as they follow the convention in Fig. 2.2

Note that $J \in \mathbb{R}^{3 \times 3}$ has the possibility of being full rank. Not all square matrices are full rank, and in this case it is full rank except for those postures that lead to singularities (more on this details later). What matters at this point is that when the J matrix is full rank, it is also invertible. Meaning that we can now solve this system uniquely in the forward and inverse directions

$$\dot{\mathbf{x}} = J(\mathbf{q})\dot{\mathbf{q}} \tag{3.18}$$

$$\dot{\mathbf{q}} = J(\mathbf{q})^{-1}\dot{\mathbf{x}} \tag{3.19}$$

The consequences of having full rank extend to the other contexts in which the Jacobian has appeared

$$\tau = J(\mathbf{q})^T \mathbf{w} \tag{3.20}$$

$$\mathbf{w} = J(\mathbf{q})^{-T} \tau \tag{3.21}$$

In this case, what are the vectors τ and $\mathbf{w}$?

3.7 Additional Comments on the Jacobian and Its Properties

Equations 3.18–3.21 are critically important as they establish the relationships of energy balance (i.e., work) between internal and external DOFs. They encapsulate the laws of conservation of energy. These different uses of the Jacobian have many levels of meaning that we will visit as this book progresses.

We have already seen that Eqs. 3.20 and 3.21 are an encapsulated way to express the equations of static equilibrium. That is, if one were to apply a force or a wrench to the endpoint of the limb, Eqs. 3.9 and 3.11 would specify the net joint torques needed to maintain the posture of the limb (i.e., satisfy the equations of static equilibrium). See Fig. 3.2.

Another important consideration that we touched upon above is that arriving at Eqs. 3.19 and 3.21 requires that the Jacobian be invertible. This is not always the case as we saw in Eq. 2.31. For the Jacobian to be invertible, it has to be a square matrix of full rank; meaning that the dimensions of the output $\mathbf{x}$ are linearly independent from each other, and equal in number to the number of independent DOFs $\mathbf{q}$. And likewise for the elements of $\mathbf{w}$ and the torques τ acting on all DOFs.

One needs to be very clear on the definition of the forward kinematic model to make sure that Eqs. 3.18–3.21 are implemented correctly and consider only independent DOFs. In essence, the issues discussed in Sect. 3.5 carry over to the definition of the Jacobian of the limb. If the Jacobian is not full rank, or invertible, then it likely means that there is an imbalance between the internal and external work the limb can do, in violation of the derivation in Sect. 3.1.

Case in point is the limb in Sect. 2.7, which has the following equations associated with it,

$$\mathbf{x} = \begin{pmatrix} x \\ y \\ \alpha \end{pmatrix} = \begin{pmatrix} G_x(\mathbf{q}) \\ G_y(\mathbf{q}) \\ G_\alpha(\mathbf{q}) \end{pmatrix} = \begin{pmatrix} l_1 c_1 + l_2 c_{12} \\ l_1 s_1 + l_2 s_{12} \\ q_1 + q_2 \end{pmatrix} \tag{3.22}$$

which produces the non square Jacobian $J \in \mathbb{R}^{3 \times 2}$

$$J = \begin{bmatrix} -l_1 s_1 - l_2 s_{12} & -l_2 s_{12} \\ l_1 c_1 + l_2 c_{12} & l_2 c_{12} \\ 1 & 1 \end{bmatrix} \tag{3.23}$$

You can see how the choice of making the Jacobian square and full rank was done by removing the third row of the forward kinematic model in Eq. 3.22 to produce Eq. 3.13. We did this because we knew from the start that the DOF α of the endpoint is not independently controllable as per our choice of forward kinematic model. The calculation of the angle α, its angular velocity, and torques at the endpoint need to be done off-line as they are not part of the system of equations associated with the full rank 2×2 Jacobian. Implicitly, we are saying that we need not calculate them because they are unimportant to us, or our analysis—and we are willing to live with the consequences of this uncontrolled and unmonitored *dependent* DOF of the endpoint (as opposed to the independently controlled DOFs of the endpoint). Therefore, you should work with Jacobian matrices that come from geometric and forward kinematic models that obey conservation of energy between internal and external work.

As a brief aside, how does one find mechanically sound solutions numerically when the Jacobian is, say, a 3×5 matrix (i.e., $J \in \mathbb{R}^{3 \times 5}$), and thus not uniquely invertible? Here any reasonable engineer would consider the use of the *Moore-Penrose pseudoinverse* to implement the different forms of the Jacobian in Eqs. 3.9–3.12. The Moore-Penrose pseudoinverse provides a least-squares error approximation to the inverse of a matrix [7]. But the pseudoinverse should be used with care as it will inherently carry some departures from conservation of energy.

You should not think of a Jacobian matrix (or any other matrix we see in this book) as simply an array of numbers, but rather as having a mechanical meaning, such as being filters that transform energy from one space of neuromechanical variables to another [8]. In the case of the Jacobian, we have the mechanical transformation of energy provided by the joint torques into work done by the endpoint wrenches, and vice versa (as per the law of conservation of energy used to derive the Jacobian 3.3). In addition, multiple cases arise in which the forward kinematic model, and its associated Jacobian, can include one or more rotational components. Then the contact boundary conditions between the endpoint of the limb and the mechanical environment become crucial to the analysis. For an example see [9].

Thus matrices are entities that, when interpreted neuromechanically, have many powerful and interesting properties such as gains, preferred input and output directions, singularities, etc. Several of these issues will be explored in later chapters, but it is highly recommended that the reader develop an intuition for the properties of full-rank and rank-deficient matrices. One motivation to those interested in neuromechanics is that the neuromuscular system of vertebrates evolved in this mechanical context of non-invertible transformations. Developing intuition with these apparently simple matrices has much to teach us about neuromuscular function.

3.8 Exercises and Computer Code

Exercises and computer code for this chapter in various languages can be found at http://extras.springer.com or found by searching the World Wide Web by title and author.

References

1. T. Yoshikawa, *Foundations of Robotics: Analysis and Control* (MIT Press, Cambridge, 1990)
2. R.M. Murray, Z. Li, S.S. Sastry, *A Mathematical Introduction to Robotic Manipulation* (CRC, Boca Raton, 1994)
3. B. Siciliano, O. Khatib, *Springer Handbook of Robotics* (Springer, Berlin, 2008)
4. G.T. Yamaguchi, *Dynamic Modelling of Musculoskeletal Motion: A Vectorized Approach for Biomechanical Analysis in Three Dimensions* (Kluwer Academic Publishers, Boston, 2001)
5. T.R. Kane, D.A. Levinson, *Dynamics, Theory and Applications* (McGraw Hill, New York, 1985)
6. F.C. Moon, *Applied Dynamics: with Applications to Multibody and Mechatronic Systems* (Wiley, New York, 2008)
7. G. Strang, *Introduction to Linear Algebra* (Wellesley Cambridge Press, Wellesley, 2003)
8. F.J. Valero-Cuevas, A mathematical approach to the mechanical capabilities of limbs and fingers. Adv. Exp. Med. Biol. **629**, 619–633 (2009)
9. F.J. Valero-Cuevas, F.E. Zajac, C.G. Burgar, Large index-fingertip forces are produced by subject-independent patterns of muscle excitation. J. Biomech. **31**, 693–703 (1998)

Chapter 4
Tendon-Driven Limbs

Abstract The purpose of this chapter is to introduce you to the fundamentals of
tendon-driven limbs, and to begin to explore how they affect our understanding of
vertebrate and robotic limbs. Many robotic limbs are driven by motors or pistons
that act on the kinematic degrees of freedom (DOFs, e.g., rotational joints) either via
linkages, cables, or gears. These actuators can exert forces and torques in both clock-
wise and counterclockwise directions, symmetrically in either direction—which in
the robotics literature are idealized and analyzed as *torque-driven limbs*. The term
'tendon-driven' comes from the robotics literature where limbs are actuated via a
variety of motors or muscles that pull on strings, cables, or tendons that cross the kinematic DOFs. Thus, these actuators can only pull on, or resist stretch in, the tendons.
But they cannot not push on the tendons. This discontinuity and asymmetry in actuation makes tendon-driven limbs distinctly different from their torque-driven counterparts with symmetric actuation. While this asymmetric actuation in tendon-driven
limbs might have some mechanical disadvantages and complicate their analysis, it
can also have advantages such as being light weight, and allowing remote actuation
and flexibility of tendon routing. As the reader will discover, varying tendon routings and moment arms can enable multiple solutions for specific functional requirements, especially when size and power constraints are critical. More importantly,
because the nervous system is unavoidably confronted with the need to actuate and
control the tendon-driven limbs in vertebrates, the nuances of tendon-driven limbs
provide insights into the nature of neural control, evolutionary adaptations, disability,
and rehabilitation that is not available in the torque-driven formulation. Note that
throughout this book, I use the terms *muscle* when relating specifically to the
behavior, forces, or state of the muscle tissue, *musculotendon* when relating to issues
that involve the muscle and its tendons of origin and insertion, and *tendon* when
relating specifically to the behavior, forces, or state of the tendon of insertion of the
muscle. For most mathematical and mechanical analyses, however, the term tendon
suffices as it applies to both robots and vertebrates. When the analysis continues on
to consider muscle mechanics and its neural control, I will prefer to use the term
musculotendon.

© Springer-Verlag London 2016
F.J. Valero-Cuevas, *Fundamentals of Neuromechanics*,
Biosystems & Biorobotics 8, DOI 10.1007/978-1-4471-6747-1_4

4.1 Tendon Actuation

Figure 4.1 shows the simplest tendon-driven limb: 1 muscle acting on a planar hinge joint. The muscle is idealized as a linear actuator attached to the proximal link (or bone), which projects its tendon distally across the joint and attaches to the distal link. As discussed earlier, this is a reasonable description because many joints in vertebrate limbs can be treated, as a first approximation, as hinge joints. The muscle force $\mathbf{f}_m$ transmitted by the tendon produces a torque τ at the DOF because the tendon crosses the joint at a distance $\mathbf{r}$ from its joint center. Given that $\mathbf{f}_m$ and $\mathbf{r}$ are vectors, the torque is calculated as

$$\tau = \mathbf{r} \times \mathbf{f}_m \tag{4.1}$$

where the operator $\times$ is the cross-product of two vectors defined as

$$\mathbf{c} = \mathbf{a} \times \mathbf{b} = \|\mathbf{a}\|\,\|\mathbf{b}\|\,sin(\alpha) \tag{4.2}$$

where α is the included angle between the two vectors, and vector $\mathbf{c}$ is perpendicular to both $\mathbf{a}$ and $\mathbf{b}$ in a direction defined by the right hand rule. Therefore, a positive torque is counterclockwise on the plane.

The vector $\mathbf{r}$ is is called the *moment arm* of that muscle at that joint. It is often described for planar systems, where Eq. 4.1 is simplified to its scalar form

Fig. 4.1 Anatomy of a 1 muscle, 1 DOF tendon-driven limb. **a** The force produced by the muscle is transmitted by the tendon to induce a rotation of the joint. **b** This can be analyzed as a muscle force $\mathbf{f}_m$ acting at a distance $\mathbf{r}$ from the joint center to produce a torque τ at the joint, where $\mathbf{r}$ is the *moment arm* of that muscle at that joint

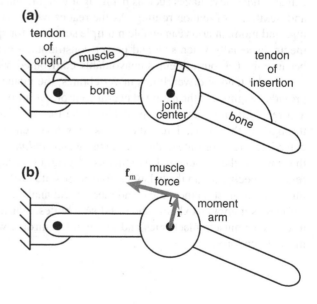

$$\|\boldsymbol{\tau}\| = \|\mathbf{r}\| \ \|\mathbf{f}_m\|$$
$$\tau = r f_m$$

(4.3)

which is easy to illustrate for joints idealized as pulleys that enforce constant or nearly constant moment arms. I will present a more formal definition in Sect. 4.3.

4.2 Tendon Routing, Skeletal Geometry, and Moment Arms

It is important to note that Eqs. 4.1 and 4.3 apply only to the instantaneous posture of the limb. That is, the moment arm of a tendon can either be constant throughout the range of motion of the joint, or be dependent on the joint angle:

- Constant moment arms
  Whenever an articulation can be modeled as a circular pulley, the scalar equation 4.3 can be used to calculate joint torque. There are precious few anatomical joints with such geometry, but this is an assumption that is often used as a first approximation in musculoskeletal models [1]. Such joints can be represented as shown in Fig. 4.1.
- Posture-dependent moment arms
  Consider the more common case where the bone contours cannot be considered a circular pulley, the moment arm is created by a system of ligamentous pulleys (Fig. 4.2), or the tendons bow-string away from the joint (Fig. 4.3). In these cases it is clear that $\mathbf{r(q)}$ needs to be calculated for each joint angle. In the case of the non-circular pulley (Fig. 4.2a), the shape of the cam defines $\mathbf{r(q)}$. In the case of

Fig. 4.2 Anatomical cases for posture of the limb-dependent moment arms

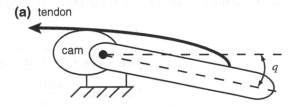

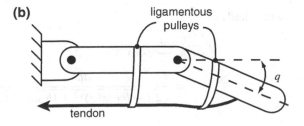

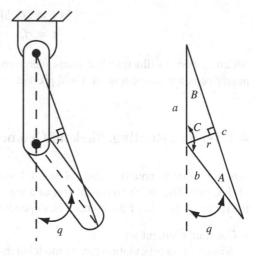

Fig. 4.3 Anatomy of posture-dependent moment arms where the tendon can bowstring away from the bones

tendon pulleys (Fig. 4.2b) or bowstringing (Fig. 4.3), it is necessary to consider the detailed geometry.

As an example, Fig. 4.3 represents an idealized biceps muscle that bowstrings as the elbow flexes. In this case, using a classical geometric analysis of triangles where the sides are lower case letters a, b, and c, and angles are upper case letters A, B, and C we have

$$\frac{r}{b} = sin(A)$$
$$\Rightarrow r = b \, sin(A) \tag{4.4}$$

Given that the DOF of the joint is the angle q defined from the vertical, we obtain

$$C = 180 - q$$
$$c = \sqrt{(a + b \, cos(q))^2 + (b \, sin(q))^2}$$
$$\text{Law of sines:} \quad \frac{sin(C)}{c} = \frac{sin(A)}{a} \tag{4.5}$$
$$\Rightarrow sin(A) = \frac{a}{c} \, sin(C)$$

which leads to

$$r = \frac{ab}{c} \, sin(C)$$
$$\Rightarrow r = \frac{ab}{\sqrt{(a + b \, cos(q))^2 + (b \, sin(q))^2}} \, sin(180 - q) \tag{4.6}$$

4.3 Tendon Excursion

The geometry of tendon actuation also has important implications to muscle force production. This comes from the fact that the force a muscle can produce depends on its length and velocity [2, 3]. Thus it is important to calculate the length and velocity of a muscle for any posture of the limb.

Consider Fig. 4.6 where we see that the change in angle δq induces an excursion, or travel, of the tendon. Call this *tendon excursion* δs. The question is, what is δs as a function of δq? We start with the simple example of a circular cam that produces a constant moment arm. Figure 4.4 shows from first principles that

$$\text{Circumference} = 2\pi r \qquad (4.7)$$

$$\delta s = r\delta q \qquad (4.8)$$

$$r = \frac{\delta s}{\delta q} \qquad (4.9)$$

These equations imply several important facts:

- If δq is in radians, for a tendon with a constant moment arm r, then $\delta s = r\delta q$ in the units of the moment arm.
- The instantaneous moment arm, $r(q)$, is equal to the partial derivative of the measured excursion with respect to the measured angle, Fig. 4.5. This is the formal definition of the moment arm $r(q)$. This geometric definition is used experimentally to extract moment arm values for complex tendon paths [4, 5].
- It is critical that a sign convention be defined to specify positive and negative excursions. I used a convention based on the definition of a positive rotation as per the right hand rule. See Fig. 4.6 and Sect. 4.6.

Fig. 4.4 Definition of excursion δs as function of r and δq as per Eq. 4.8

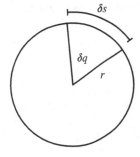

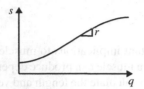 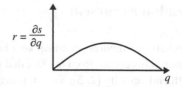

Fig. 4.5 Schematic representation of the calculations of the moment arm, $r(q)$. Given tendon excursion as a function of joint angle $s(q)$, the moment arm is its partial derivative with respect to the joint angle $r(q) = \frac{ds}{dq}$, Eq. 4.9 [4]

Fig. 4.6 Measurement of tendon excursion for a simple tendon path. Note that a negative (as per the right hand rule) rotation of the joint $-\delta q$ induces a positive (rightward) tendon excursion δs that lengthens the musculotendon, and vice versa. See Sect. 4.6 for a definition of this sign covention

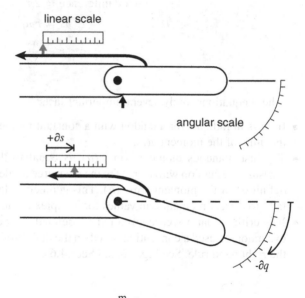

Fig. 4.7 A planar 1 DOF limb driven by 6 tendons

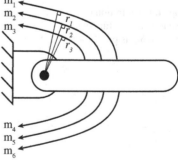

4.4 Two or More Tendons Acting on a Joint: Under- and Overdetermined Systems

Consider Fig. 4.7, where 6 muscles cross a planar joint. In this case we can sum the scalar equation 4.3 for each muscle to find the net joint torque at the joint, and also begin to use the notation of vector multiplication[1]

$$\tau = \sum_{i=1}^{6} r(q)_i \ f_i \tag{4.10}$$

$$\text{or } \tau = \left(r(q)_1, \ r(q)_2, \ r(q)_3, \ -r(q)_4, \ -r(q)_5, \ -r(q)_6\right)^T \begin{pmatrix} f_1 \\ f_2 \\ f_3 \\ f_4 \\ f_5 \\ f_6 \end{pmatrix} \tag{4.11}$$

following the conventions that forces f_i are positive values because muscles cannot push, but only pull; and moment arms $r(q)_i$ are absolute distances where their signs depend on whether their torque would induce a positive or negative rotation as per the right hand rule. Notice that τ is a scalar indicating the numerical value of the net joint torque.

More formally, for a joint crossed by N muscles

$$\tau = \sum_{i=1}^{N} r(q)_i \ f_i \tag{4.12}$$

such summation can be, in general, expressed as the dot product of the vector of scalar moment arms $\mathbf{r}(q)$ and the vector of scalar muscle forces $\mathbf{f}_m$

$$\tau = \mathbf{r}(q) \cdot \mathbf{f}_m \tag{4.13}$$

which is equivalent to

---

[1]A word about *vectors*: often the term vector is used to mean a multidimensional physical entity like a 3D location, a 3D force, etc. as was done in Chaps. 2 and 3. However, *vector algebra* applies to any situation where several items are grouped for the purposes of mathematical operations. Case in point, assembling a vector of muscle forces f_i, or a vector of moment arm values $r(q)_i$ is useful for the mathematical operation of calculating net joint torques. However, as described in Part III, the concept of *vector space* of muscle forces, etc., has the deeper meaning of describing high-dimensional spaces of particular variables [6]. See Appendix A.

$$\tau = \mathbf{r}(q)^T \mathbf{f}_m \qquad (4.14)$$

and

$$\tau = \begin{pmatrix} r(q)_1 \\ r(q)_2 \\ \vdots \\ r(q)_N \end{pmatrix} \cdot \begin{pmatrix} f_1 \\ f_2 \\ \vdots \\ f_N \end{pmatrix} = \left(r(q)_1, \ r(q)_2, \ \dots, \ r(q)_N \right)^T \begin{pmatrix} f_1 \\ f_2 \\ \vdots \\ f_N \end{pmatrix} \qquad (4.15)$$

This represents a mapping from the *vector space* of muscle forces (of dimension N, or $\mathbf{f}_m \in \mathbb{R}^N$) to the scalar value of net joint torque (of dimension 1, or $\tau \in \mathbb{R}^1$),

$$\mathbb{R}^N \to \mathbb{R}^1 \qquad (4.16)$$

Equations 4.12–4.15 are *underdetermined* in that they equivalently express that a given value of the net joint torque scalar can be produced by a variety of combinations of muscle forces. Essentially, there are N muscle activation DOFs that can be combined in an infinite number of ways to achieve a same goal.[2] This is a form of *muscle redundancy* that begs the question of how the nervous system selects a solution from among many. This has been called the central problem of motor control [7]. We will discuss this in detail in Chap. 5.

An equally important concept described by Fig. 4.7—which is not as well highlighted in the literature [8]—concerns the tendon excursions of the N muscles that cross the joint. In this case of a single DOF q

$$\delta s_i = -r(q)_i \delta q \quad \text{for } i = 1, \dots, N \qquad (4.17)$$

$$\begin{pmatrix} \delta s_1 \\ \delta s_2 \\ \vdots \\ \delta s_N \end{pmatrix} = \begin{pmatrix} r(q)_1 \\ r(q)_2 \\ \vdots \\ r(q)_N \end{pmatrix} \delta q \qquad (4.18)$$

which in vector form is

$$\delta \mathbf{s} = -\mathbf{r}(q) \, \delta q \qquad (4.19)$$

---

[2]Note that the letter M need not stand for muscles, nor N be used only for kinematic DOFs of a limb. They are simply letters to indicate indices and dimensions. See Appendix A.

The negative sign before the moment arm vector $\mathbf{r}(q)$ comes from the sign convention shown in Fig. 4.6. This vector multiplication represents a mapping from the scalar value of change in joint angle (of dimension 1, or $q \in \mathbb{R}^1$), to the high-dimensional vector space of tendon excursions (of dimension N, or $\delta \mathbf{s} \in \mathbb{R}^N$).

$$\mathbb{R}^1 \rightarrow \mathbb{R}^N \tag{4.20}$$

Equations 4.17 and 4.19 are *overdetermined* in that they equivalently express that a change in a single DOF δq determines the excursions δs of all N tendons. Said differently, if for some reason any of the muscles fail to stretch to keep up with the joint rotation, the single joint will lock up. This is the opposite of redundancy—and begs the question of how the nervous system controls the excursions of all muscles so that the limb can move smoothly. We have raised this issue in [8], and we will discuss this in detail in Chap. 6.

4.5 The Moment Arm Matrix for Torque Production

The concepts of moment arms and tendon excursions presented above extend to the case of multiple muscles crossing multiple joints. Consider the planar system with 2 DOFs and 4 muscles (tendons or musculotendons) shown in Fig. 4.8.

By convention, the moment arms of this limb with M DOFs and N muscles can be compiled as a *moment arm matrix*, $R(\mathbf{q}) \in \mathbb{R}^{M \times N}$, with M rows and N columns, Eq. 4.21. The entries of the $R(\mathbf{q})$ matrix are the $r(q)_{i,j}$ values, which are signed scalars (as per the convention shown in Fig. 4.6), i is the index indicating the DOF and ranges from 1 to M, and j is the index indicating the tendon (or muscle or musculotendon) and ranges from 1 to N.

$$R(\mathbf{q}) = \begin{bmatrix} r(\mathbf{q})_{1,1} & r(\mathbf{q})_{1,2} & r(\mathbf{q})_{1,3} & \cdots & \cdots & r(\mathbf{q})_{1,N} \\ r(\mathbf{q})_{2,1} & r(\mathbf{q})_{2,2} & r(\mathbf{q})_{2,3} & \cdots & \cdots & r(\mathbf{q})_{2,N} \\ \vdots & \vdots & \vdots & \vdots & \vdots & \vdots \\ r(\mathbf{q})_{M,1} & r(\mathbf{q})_{M,2} & r(\mathbf{q})_{M,3} & \cdots & \cdots & r(\mathbf{q})_{M,N} \end{bmatrix} \tag{4.21}$$

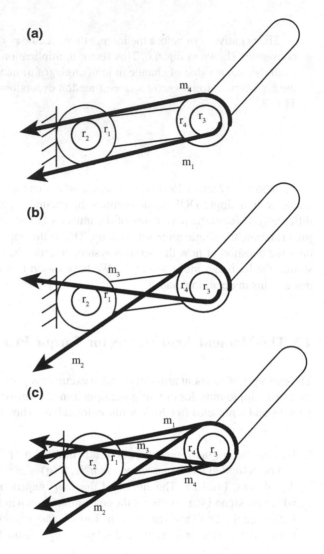

The moment arm matrix can be used to transform the vector of muscle forces $\mathbf{f}_m$ transmitted by their tendons into the vector of joint torques $\boldsymbol{\tau}$ using the following vector-matrix equation

$$\begin{pmatrix} \tau_1 \\ \tau_2 \\ \vdots \\ \tau_M \end{pmatrix} = R(\mathbf{q}) \begin{pmatrix} f_1 \\ f_2 \\ \vdots \\ f_N \end{pmatrix} \tag{4.22}$$

or

$$\tau = R(\mathbf{q})\,\mathbf{f}_m \tag{4.23}$$

Note that, because there are usually more muscles than kinematic DOFs

$$R(\mathbf{q}) \in \mathbb{R}^{M \times N} \tag{4.24}$$

$$\text{where } M < N$$

meaning that there are *more columns than rows*:

- The moment arm matrix is a wide, non-square matrix, characteristic of an *underdetermined system* (i.e., more variables than equations)
- $R(\mathbf{q})$ is *noninvertible*

Therefore, Eq. 4.23 is a generalized version of Eqs. 4.12–4.15, making them underdetermined or *redundant* systems. What is characteristic of underdetermined systems is that they have at least one solution, but usually many more. That is, a given set of net joint torques (a *net joint torque vector* τ) can be produced by an infinite variety of combinations of muscle forces (or *muscle force vectors* $\mathbf{f}_m$). This is a *forward model* calculation because it follows the natural sequence of events: the inputs are the muscle forces delivered to the tendons, and the outputs are the net joint torques that arise. Again, this is the definition of *muscle redundancy*. These kinds of underdetermined problems are solved in engineering by using *numerical optimization*. Numerical optimization uses a *cost function* to evaluate the relative merit of each possible solution, and then tries to find the unique best possible solution as per that criterion [9–11]. That is, knowing that there are many solutions, then the best solution is the one that maximizes or minimizes the cost function. See Sect. 5.2. This optimization approach to solving such underdetermined systems is the dominant approach, and the classical interpretation of muscle redundancy, in neuromuscular control [8].

As an example, the moment arm matrix for the limb in Fig. 4.8, $R(\mathbf{q})$ is

$$R(\mathbf{q}) = \begin{bmatrix} -r_1 & -r_1 & r_2 & r_2 \\ -r_3 & r_4 & -r_3 & r_4 \end{bmatrix} \tag{4.25}$$

Two other examples of robotic fingers and their associated moment arm matrices are shown in Fig. 4.9. They have, respectively, 4 DOFs and 5 muscles, and 4 DOFs and 8 muscles.

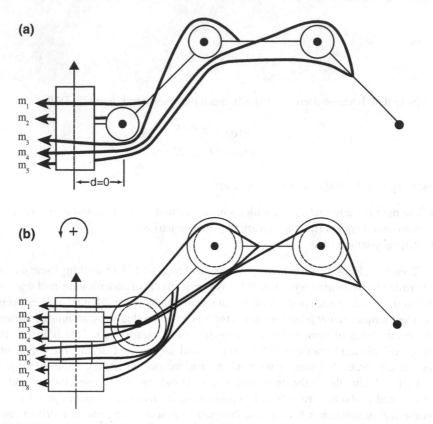

Fig. 4.9 Moment arm matrices for two sample limbs. **a** A 4 DOF and 5-muscle limb, and **b** a 4 DOF and 8-muscle limb. They each have a particular tendon routing to actuate the 4 DOFs. In both figures, the first joint is a universal joint with 2 overlapping DOFs—which we show as separated in the figure for clarity, hence $d = 0$. The first DOF (its axis of rotation points upwards) allows these limbs to move side-to-side in 3D: into the plane (for a positive rotation as per the right hand rule) and out of the plane (for a negative rotation). The other 3 DOFs are flexion-extension with their axes of rotation coming out of the page, thus a positive rotation is counterclockwise as per the right hand rule. The systematic way in which a moment arm matrix is constructed is as follows. Take muscle 1 in (**a**) as an example. It crosses joints 1 and 2, and its moment arm values are contained in the first column of its R matrix. It will produce a negative rotation or torque around the first joint, so the entry in the first row is -1. It will produce a positive rotation or torque around the second joint, so the entry in the second row is $+1$. It does not cross the other joints, so the moment arm value at those joints is zero. The rest of the R matrix is similarly constructed. Adapted with permission from [12]

4.6 The Moment Arm Matrix for Tendon Excursions

Let us now look at the generalized versions of Eqs. 4.17 and 4.19 for multi-joint, multi-muscle limbs in vector-matrix form. Consider the planar system with 2 DOFs and 4 tendons shown in Fig. 4.8.

Once again, by convention, the moment arm values are entries of a matrix $r(q)_{i,j}$, where, i is the joint number and ranges from 1 to M, and j is the tendon (or muscle or musculotendon) number, which ranges from 1 to N, as in Eq. 4.21. By convention, a moment arm is defined with positive values indicating positive rotation or torque generated at a joint when tension is applied to the tendon, and vice versa. However, in this case we must consider that when a muscle induces a positive joint rotation or produces a positive torque, it must shorten or try to shorten as shown in Fig. 4.6. Therefore, a positive rotation with a *positive moment arm* induces a *musculotendon shortening* and therefore a *negative tendon excursion*—and vice versa.

Thus if the entries $r(q)_{i,j}$ are arranged to represent the effect of a positive rotation of joint i on the excursion of tendon j, we obtain

$$
\begin{pmatrix} \delta s_1 \\ \delta s_2 \\ \delta s_3 \\ \vdots \\ \vdots \\ \delta s_N \end{pmatrix} = \begin{bmatrix} -r(\mathbf{q})_{1,1} & -r(\mathbf{q})_{2,1} & \cdots & -r(\mathbf{q})_{M,1} \\ -r(\mathbf{q})_{1,2} & -r(\mathbf{q})_{2,2} & \cdots & -r(\mathbf{q})_{M,2} \\ -r(\mathbf{q})_{1,3} & -r(\mathbf{q})_{2,3} & \cdots & -r(\mathbf{q})_{M,3} \\ \vdots & \vdots & \vdots & \vdots \\ \vdots & \vdots & \vdots & \vdots \\ -r(\mathbf{q})_{1,N} & -r(\mathbf{q})_{2,N} & \cdots & -r(\mathbf{q})_{M,N} \end{bmatrix} \begin{pmatrix} \delta q_1 \\ \delta q_2 \\ \vdots \\ \delta q_M \end{pmatrix}
$$

(4.26)

Comparing Eqs. 4.21 and 4.26, note that the individual entries $r(q)_{i,j}$ are placed in an arrangement that creates the negative of the transpose of the moment arm matrix: $-R(\mathbf{q})^T$. This matrix transforms a vector of joint rotations $\delta \mathbf{q}$ into a vector of tendon excursions $\delta \mathbf{s}$ using the following equation

$$\delta \mathbf{s} = -R(\mathbf{q})^T \delta \mathbf{q}$$

(4.27)

where you should notice the *negative sign* before $R(\mathbf{q})^T$. Because there are usually more muscles than joints, it means that there are *more rows than columns*,

$$R(\mathbf{q})^T \in \mathbb{R}^{N \times M}$$

(4.28)

and still $M < N$

This means that:

- The transpose of the moment arm matrix is a *tall*, non-square matrix, characteristic of an *overdetermined system* (i.e., fewer variables than equations)
- $R(\mathbf{q})^T$ is not invertible

Therefore, Eqs. 4.27 and 4.28 represent an overdetermined relationship between joint rotations and tendon excursions. What is characteristic of overdetermined systems is that there can be at most one solution, if it exists. But often there is no solution. That is, if you rotate the joints, the tendons will undergo excursions—*if and only if* all the muscles that are being stretched are allowed to undergo the necessary excursions to accommodate the joint rotations. The rotation of a few joints determines the length changes of *all* musculotendons. These kinds of overdetermined problems are solved in engineering by finding the least squares error solution [10]. That is, knowing that there is at most one solution—and likely no solution because you are trying to meet many constraints with not enough variables—then the best solution is the one that violates the constraints as little as possible. These *least bad* solutions are the ones that minimize the square of the error, hence the name *least squares* error.

4.7 Implications to the Neural Control of Tendon-Driven Limbs

As mentioned above, overdetermined systems have at most one solution. In this case, the one valid solution is when all musculotendons being stretched can undergo the required length changes. Here we confront once again the asymmetry of tendon-driven limbs where a tendon can be over-shortened by simply going slack, but can only be 'over-lengthened' if it is possible to stretch it. Consider what would happen if, to achieve a given joint rotation, those muscles that need to stretch are not able to undergo the necessary length change. Physiologically, this could be due to mechanical limitations (like when you cannot flex your middle finger while the other fingers remain straight—the bane of musicians the world over [13]), neural coupling among muscles [14], or neural resistance to lengthening due to *stretch reflexes*. These stretch reflexes are neuronal circuits in the spinal cord that naturally force muscles to resist being stretched by external forces as a means to maintain limb posture in the presence of perturbations. As far back as the work of the famed neuroscientist Sir Charles Sherrington, it has been recognized that natural and versatile limb function requires both activation and inhibition of muscles [15].

In the current literature, this problem of overdetermined solutions for joint rotations is not very well understood or developed, as discussed in [8, 16]. But it is recognized that certain disruptions of the stretch reflexes lead to disruptions of limb movement—as we would expect if muscles that need to lengthen to accommodate the motion are not able to do so as per Eq. 4.27. Section 6.4 delves into these neuro-mechanical interactions, and their clinical consequences.

4.8 Exercises and Computer Code

Exercises and computer code for this chapter in various languages can be found at http://extras.springer.com or found by searching the World Wide Web by title and author.

References

1. F.J. Valero-Cuevas, H. Hoffmann, M.U. Kurse, J.J. Kutch, E.A. Theodorou, Computational models for neuromuscular function. IEEE Rev. Biomed. Eng. **2**, 110–135 (2009)
2. F.E. Zajac, Muscle and tendon: properties, models, scaling, and application to biomechanics and motor control. Crit. Rev. Biomed. Eng. **17**(4), 359–411 (1989)
3. F.E. Zajac, How musculotendon architecture and joint geometry affect the capacity of muscles to move and exert force on objects: a review with application to arm and forearm tendon transfer design. J. Am. Hand Surg. **17**(5), 799–804 (1992)
4. K.N. An, Y. Ueba, E.Y. Chao, W.P. Cooney, R.L. Linscheid, Tendon excursion and moment arm of index finger muscles. J. Biomech. **16**(6), 419–425 (1983)
5. M.U. Kurse, H. Lipson, F.J. Valero-Cuevas, Extrapolatable analytical functions for tendon excursions and moment arms from sparse datasets. IEEE Trans. Biomed. Eng. **59**(6), 1572–1582 (2012)
6. F.J. Valero-Cuevas, A mathematical approach to the mechanical capabilities of limbs and fingers. Adv. Exp. Med. Biol. **629**, 619–633 (2009)
7. N.A. Bernstein, *The Co-ordination and Regulation of Movements* (Pergamon Press, New York, 1967)
8. F.J. Valero-Cuevas, B.A. Cohn, H.F. Yngvason, E.L. Lawrence, Exploring the high-dimensional structure of muscle redundancy via subject-specific and generic musculoskeletal models. J. Biomech. **48**(11), 2887–2896 (2015)
9. V. Chvatal, *Linear Programming* (W.H. Freeman and Company, New York, 1983)
10. G. Strang, *Introduction to Linear Algebra* (Wellesley Cambridge Press, Wellesley, 2003)
11. P.E. Gill, W. Murray, M.H. Wright, *Practical Optimization* (Academic Press, New York, 1981)
12. J.M. Inouye, J.J. Kutch, F.J. Valero-Cuevas, A novel synthesis of computational approaches enables optimization of grasp quality of tendon-driven hands. IEEE Trans. Robot. **28**(4), 958–966 (2012)
13. R.A. Henson, H. Urich, Schumann's hand injury. Br. Med. J. **1**(6117), 900 (1978)
14. M.H. Schieber, M. Santello, Hand function: peripheral and central constraints on performance. J. Appl. Physiol. **96**(6), 2293–2300 (2004)
15. C.S. Sherrington, Reflex inhibition as a factor in the co-ordination of movements and postures. Exp. Physiol. **6**(3), 251–310 (1913)
16. J.J. Kutch, F.J. Valero-Cuevas, Challenges and new approaches to proving the existence of muscle synergies of neural origin. PLoS Comput. Biol. **8**(5), e1002434 (2012)

Part II
Introduction to the Neural Control
of Tendon-Driven Limbs

Neuromechanics requires an understanding of the relationship between the number of muscles and the number of kinematic degrees of freedom. The first chapter of this Part introduces the reader to fundamental concepts associated with the classical view of muscle redundancy. In this underdetermined phrasing of the problem of motor control, there are many solutions to how muscles must coordinate to produce a particular mechanical task. However, when addressing tendon excursions—which only arises when treating limbs as tendon-driven—we are confronted with an overdetermined system. I develop this idea in the second Chapter of this Part. Taken together these Chapters lay the foundation for Part III where I introduce and develop the concept of feasible neural commands.

Chapter 5
The Neural Control of Joint Torques in Tendon-Driven Limbs Is Underdetermined

Abstract This chapter introduces the mathematical foundations of the classical notion of muscle redundancy. As presented in Chap. 4, a sub-maximal net torque at a joint actuated by tendons can be produced by a variety of combinations of individual forces at each tendon. We see this already in the simplest case of a planar joint with 2 tendons—one on each side of the joint. Of course, each combination of tendon forces will produce different loading at the tendons and joint, and will incur different metabolic or energetic costs, etc. But in principle there are multiple solutions to the problem of achieving a given mechanical output. This underdetermined problem is called the problem of *muscle redundancy*, and it begs the question of how the nervous system (or a robotic controller) should select a particular solution from among many. This has been called the central problem of motor control and has occupied much of the literature in this field. The main goal of this chapter, however, is to introduce and cast this problem for high-dimensional multi-joint, multi-muscle limbs (it is often only presented in simplified joints). This will serve as the foundation of subsequent chapters where we critically assess this classical notion of muscle redundancy— and challenge its assumptions and conclusions. As mentioned in Chap. 1, however valuable and informative the concept of muscle redundancy has been, it is also paradoxical with respect to the evolutionary process and clinical reality, and should be revised.

5.1 Muscle Activation and Redundancy of Neural Control

Having seen the kinematics and mechanics of torque- and tendon-driven limbs, we can now add muscle forces to complete our model of how neural commands produce motor output. But first, what is a neural command? In our fields, the terms *muscle activation*, *muscle excitation*, *neural command*, *neural drive*, *motor command*, *excitation level*, and others are used (often interchangeably) to quantify the extent to which the nervous system is delivering excitatory impulses to the muscle tissue. This is a complex topic that can be seen from multiple perspectives and to varying degrees of nuance as described in, say, [1]. But for our purposes, I take the modeling

© Springer-Verlag London 2016   55
F.J. Valero-Cuevas, *Fundamentals of Neuromechanics*,
Biosystems & Biorobotics 8, DOI 10.1007/978-1-4471-6747-1_5

perspective [2] that the nervous system issues a control command, $a_i(t)$ at time t, which I call *muscle activation* in this book.

By convention, muscle activation is a number assumed to be bounded between 0 and 1, where its normalized value represents the range between zero and maximal neural activation, respectively. This convention is the formal expression of the fact that muscles can only pull voluntarily, but not push. That is, the nervous system can only command a muscle to actively shorten (or try to shorten or resist stretch) to produce tension at its tendon. But the nervous system cannot command a muscle to lengthen and thereby push on its tendon. If there are multiple muscles, then the vector of muscle activations $\mathbf{a}(t)$ contains the control commands to all muscles at that point in time. For simplicity and without loss of generalization, however, I will only use the time variable when necessary, and define the vector $\mathbf{a}$ for N muscles as follows:

$$\mathbf{a} = \begin{pmatrix} a_1 \\ a_2 \\ \vdots \\ a_N \end{pmatrix} \tag{5.1}$$

where $0 \le a_i \le 1$ for $i = 1, \ldots, N$

The activation of a muscle leads to the production of the *muscle force*, $f_i(t)$, considered in Eq. 4.12. Once again, in our field we use this term synonymously with *tendon tension, tendon force, musculotendon force*, etc. When grouping together muscle forces, we can speak of the vector $\mathbf{f}_m$ used in Eq. 4.13, again dropping the time variable for simplicity

$$\mathbf{f}_m = \begin{pmatrix} f_1 \\ f_2 \\ \vdots \\ f_N \end{pmatrix} \tag{5.2}$$

where $0 \le f_i \le F_{0i}$ for $i = 1, \ldots, N$

where F_{0i} is the scalar value of the maximal force muscle i can produce. F_{0i} depends on muscle architecture, like physiological cross sectional area and pennation angle, as well as the current muscle fiber length and velocity [3] Given that muscle fiber length and velocity depend on tendon excursion and its time derivative, respectively, then F_{0i} is best written as $F_{0i}(\mathbf{q}, \dot{\mathbf{q}})$ (time is implicitly considered).

To obtain the vector of muscle forces $\mathbf{f}_m$, one can multiply the vector of muscle activations, $\mathbf{a}$, by a matrix F_0 containing the N scalar values along its diagonal

$$F_0(\mathbf{q}, \dot{\mathbf{q}}) = \begin{bmatrix} F_{01} & & & \mathbf{0} \\ & F_{02} & & \\ & & \ddots & \\ \mathbf{0} & & & F_{0N} \end{bmatrix} \qquad (5.3)$$

Therefore

$$\mathbf{f}_m = F_0(\mathbf{q}, \dot{\mathbf{q}})\,\mathbf{a} \qquad (5.4)$$
$$\text{and}\quad \boldsymbol{\tau} = R(\mathbf{q})\,F_0(\mathbf{q}, \dot{\mathbf{q}})\,\mathbf{a} \qquad (5.5)$$

Combining Eqs. 3.21, 4.23, and 5.4 [4] we obtain the complete mapping from muscle activations, $\mathbf{a}$, to endpoint wrenches, $\mathbf{w}$, via:

- The force production capabilities of muscles, $F_0(\mathbf{q}, \dot{\mathbf{q}})$, which transform muscle activations into muscle forces $\mathbf{f}_m$
- The moment arm matrix, $R(\mathbf{q})$, which transforms muscle forces into net joint torques $\boldsymbol{\tau}$
- And the Jacobian inverse transpose of the limb, $J(\mathbf{q})^{-T}$, which transforms net joint torques into endpoint wrenches $\mathbf{w}$

stated as

$$\mathbf{w} = J(\mathbf{q})^{-T}\,R(\mathbf{q})\,F_0(\mathbf{q}, \dot{\mathbf{q}})\,\mathbf{a} \qquad (5.6)$$

For the static case (i.e., $\dot{\mathbf{q}} = 0$) at a given posture (i.e., fixed and known $\mathbf{q}$) this can be written more simply as

$$\mathbf{w} = J^{-T}\,R\,F_0\,\mathbf{a} \qquad (5.7)$$

which avoids cumbersome notation whenever dependencies on time, joint angles, and angular velocities are given to be understood, or explicitly stated as needed.

Equations 5.6 and 5.7 are systems of underdetermined linear equations where multiple neural commands can produce the same output wrench. These equations remain linear only at a given posture as, at the very least, the Jacobian will change with a change of posture.

For realistic limbs, the number of muscles, N, will always be greater than the number of kinematic DOFs. For example,

For a full planar limb with 3 DOFs, the output vector $\mathbf{w}$ exists in $\mathbb{R}^3$

thus the matrix $[J^{-T} \ R \ F_0] \in \mathbb{R}^{3 \times N}$ represents a mapping $\mathbb{R}^N \to \mathbb{R}^3$

$$(5.8)$$

And for a full 3D limb with 6 DOFs, the output vector $\mathbf{w}$ exists in $\mathbb{R}^6$

thus the matrix $[J^{-T} \ R \ F_0] \in \mathbb{R}^{6 \times N}$ represents a mapping $\mathbb{R}^N \to \mathbb{R}^6$

But of course you can have, for example, a limb in 3D space that only has a wrench output of dimensionality 4 or 5, as mentioned in Sects. 2.6 and 3.5. You now see another instance of the impact of the definition of your limb's geometric model on the output wrench its endpoint can produce.

5.2 Linear Programming Applied to Tendon-Driven Limbs

Any introduction to muscle redundancy also requires an introduction to the engineering approach to solve such problems. Underdetermined problems are solved in engineering by using *numerical optimization*. Numerical optimization uses a *cost function* to evaluate the relative merit of each possible solution, and then tries to find the unique best possible solution as per that criterion [5–7]. That is, knowing that there are many solutions, the best solution is the one that maximizes or minimizes a given cost function.

Linear programming is a branch of mathematics that finds optimal solutions to problems posed as underdetermined systems of linear equations [5], that is, with a linear cost function and linear inequality constraints. In 1947, G.B. Dantzig developed the *simplex method* to solve logistics and resource allocation problems for the U.S. Air Force using linear programming [8]. It soon became clear that a surprisingly wide range of apparently unrelated problems in production management could be stated in linear programming terms, and solved by the simplex method. For example, early in his research, L.V. Kantorovich applied the analytical technique of linear programming to demonstrate how economic planning in his country could be improved. This resulted in the Royal Swedish Academy of Sciences awarding him and T.C. Koopmans the 1975 Nobel Prize in economic science for their contributions to the theory of optimal allocation of resources.

In the context of underdetermined problems in neuromechanics, the resource to allocate and optimize is muscle activation, and the constraints to meet are those of the particular task—such as the direction of an endpoint force. Therefore the goal is to find, from among the infinite possible options, the one best activation pattern that produces the desired task.

5.2.1 Canonical Formulation of the Linear Programming Problem

To use linear programming, it is important to use a standard formulation because available algorithms assume it. The *standard* or *canonical* way to pose a linear programming problem is

Given N variables x_i (i.e., $\mathbf{x} \in \mathbb{R}^N$)

$$\text{Minimize} \quad \mathbf{c}^T \mathbf{x}$$
$$\text{subject to } A\mathbf{x} \leq \mathbf{b} \tag{5.9}$$

where $\mathbf{c}^T \mathbf{x}$ is the linear cost function, and the matrix A and the vector $\mathbf{b}$ define the linear inequality constraint equations on the N variables.

We will now see how such problems are posed and solved.

5.2.2 A Classical Example of Linear Programming: The Diet Problem

A classical example, as presented by Chvátal in [5], is as follows. Polly, a fellow university student, needs to spend as little money as possible on food, while satisfying her nutritional needs. That is, she requires *at least* 2,000 kcal of energy, 55 g of protein, and 800 mg of calcium each day. Her meal plan offers 6 food choices (yes, quite dreadful), and she has set for herself a limit for the maximal servings per day for each food choice.

As shown in Table 5.1, each food choice is one of 6 independent variables $x_1, x_2, \ldots, x_6$. Those of us who have eaten at restaurants know intuitively that there are infinitely many possible combinations of food options, but hopefully one that minimizes cost while meeting our nutritional needs (i.e., that this is an underdetermined problem). But, how do we know that the diet problem is truly underdetermined? Optimization problems can be of several types:

- Those with a unique optimal solution
- Those with no solution (infeasible)
- Those with many solutions with the same cost
- Those with infinitely many different solutions with different costs (unbounded).

For simple underdetermined linear problems such as this one, there is very likely one unique optimal solution if it is well posed. We begin by stating the problem as one

Table 5.1 Variables and details for the diet problem

Variable	Food choice	Energy	Protein	Calcium	$\frac{Cents}{Serving}$	$\frac{Max\ servings}{Day}$
x_1	Oatmeal	110	4	2	3	4
x_2	Chicken	205	32	12	24	3
x_3	Eggs	160	13	54	13	2
x_4	Whole milk	160	8	285	9	8
x_5	Cherry pie	420	4	22	20	2
x_6	Pork and beans	260	14	80	19	2

to minimize the linear equation

$$\text{cost} = 3\,x_1 + 24\,x_2 + 13\,x_3 + 9\,x_4 + 20\,x_5 + 19\,x_6 \qquad (5.10)$$

subject to the inequality constraints on nutritional value

$$110\,x_1 + 205\,x_2 + 160\,x_3 + 160\,x_4 + 420\,x_5 + 260\,x_6 \geq 2000$$
$$4\,x_1 + 32\,x_2 + 13\,x_3 + 8\,x_4 + 4\,x_5 + 14\,x_6 \geq 55 \qquad (5.11)$$
$$2\,x_1 + 12\,x_2 + 54\,x_3 + 285\,x_4 + 22\,x_5 + 80\,x_6 \geq 800$$

and subject to the inequality constraints on the valid values for each variable

$$0 \leq x_1 \leq 4$$
$$0 \leq x_2 \leq 3$$
$$0 \leq x_3 \leq 2$$
$$0 \leq x_4 \leq 8 \qquad (5.12)$$
$$0 \leq x_5 \leq 2$$
$$0 \leq x_6 \leq 2$$

To express this problem in canonical form we define the vectors $\mathbf{c}$ and $\mathbf{b}$, and the matrix A, as per Eq. 5.9. Most optimization software will solve linear equations (like MATLAB's `linprog` command) only if phrased using this convention.

$$c^T x = (3,\ 24,\ 13,\ 9,\ 20,\ 19)\,x \qquad (5.13)$$

$$
Ax =
\begin{bmatrix}
-110 & -205 & -160 & -160 & -420 & -260 \\
-4 & -32 & -13 & -8 & -4 & -14 \\
-2 & -12 & -54 & -285 & -22 & -80 \\
\hline
1 & 0 & 0 & 0 & 0 & 0 \\
0 & 1 & 0 & 0 & 0 & 0 \\
0 & 0 & 1 & 0 & 0 & 0 \\
0 & 0 & 0 & 1 & 0 & 0 \\
0 & 0 & 0 & 0 & 1 & 0 \\
0 & 0 & 0 & 0 & 0 & 1 \\
\hline
-1 & 0 & 0 & 0 & 0 & 0 \\
0 & -1 & 0 & 0 & 0 & 0 \\
0 & 0 & -1 & 0 & 0 & 0 \\
0 & 0 & 0 & -1 & 0 & 0 \\
0 & 0 & 0 & 0 & -1 & 0 \\
0 & 0 & 0 & 0 & 0 & -1
\end{bmatrix}
x \le
\begin{pmatrix}
-2000 \\
-55 \\
-800 \\
4 \\
3 \\
2 \\
8 \\
2 \\
2 \\
0 \\
0 \\
0 \\
0 \\
0 \\
0
\end{pmatrix}
= b \qquad (5.14)
$$

Let us look at Eqs. 5.13 and 5.14 carefully. First you will note that 5.13 corresponds to Eq. 5.10, but in vector form. Equation 5.14 corresponds to Eqs. 5.11 and 5.12, but each of the sub-matrices in matrix A requires attention. Notice the minus signs. These are necessary in the top sub-matrix of A because, as you can see in Eq. 5.9, the canonical form specifically defines the problem as one of *minimization* of the cost function. Also, the constraint equations must be a *less than or equal* inequality. You see this once again in the bottom sub-matrix of A given that those rows specify that each variable must be *greater than or equal* to zero. If you want a constraint to reflect a *greater than or equal* inequality, you need to invert the signs of both sides of the equation. That is,

$$a\,x \le b \text{ is equivalent to } -a\,x \ge -b \qquad (5.15)$$

Importantly, a minimization problem can be made into a maximization problem by simply changing the sign of the cost function (see Sect. 5.3). The middle sub-matrix represents the upper bounds of each of the variables, so it does not need any sign inversions. Notice also that the middle and bottom sub-matrices can be simply expressed using identity matrices of dimension 6 (i.e., I_6). Thus a more succinct version is

$$Ax = \begin{bmatrix} -110 & -205 & -160 & -160 & -420 & -260 \\ -4 & -32 & -13 & -8 & -4 & -14 \\ -2 & -12 & -54 & -285 & -22 & -80 \\ \hline & & & I_6 & & \\ \hline & & & -I_6 & & \end{bmatrix} \mathbf{x} \le \begin{pmatrix} -2000 \\ -55 \\ -800 \\ 4 \\ 3 \\ 2 \\ 8 \\ 2 \\ \hline 2 \\ 0 \\ 0 \\ 0 \\ 0 \\ 0 \\ 0 \end{pmatrix} = \mathbf{b} \quad (5.16)$$

Some trial and error shows that one *feasible* solution is $x_1 = 0$, $x_2 = 0$, $x_3 = 0$, $x_4 = 8$, $x_5 = 2$, $x_6 = 0$, indicating the number of servings of each food choice, with a cost of $72 + 40 = 112$ cents per day. Yes, back then cents were worth something.

However, the *optimal* solution $\mathbf{x}^*$ (pronounced x-star) is: $x_1 = 4$, $x_2 = 0$, $x_3 = 0$, $x_4 = 4.5$, $x_5 = 2$, $x_6 = 0$ or

$$\mathbf{x}^* = \begin{pmatrix} 4 \\ 0 \\ 0 \\ 4.5 \\ 2 \\ 0 \end{pmatrix} \quad (5.17)$$

with a cost of 92.5 cents per day.

But also notice that counting constraints does not obviously tell you whether the problem is underdetermined, overdetermined, has a solution, is infeasible, unbounded, etc. This problem has 6 variables and 15 constraints (i.e., the rows of the A matrix), yet it is underdetermined and has a unique solution! You could pose the problem as having fewer constraints that make it infeasible, such as requiring x_4 to be ≤ 6 and ≥ 12. Thus it is important that you develop an intuition of what each constraint means so that you can pose the problem and interpret the results. In Sect. 5.4, I begin to provide a geometric approach to build that intuition.

5.3 Linear Programming Applied to Neuromuscular Problems

We can now revisit our underdetermined problem, as phrased in its general form in Eq. 5.7. Consider a planar system with 2 DOFs and 4 muscles, as in Fig. 4.8 and Eq. 5.7

$$\mathbf{f} = \begin{pmatrix} f_x \\ f_y \end{pmatrix} = J^{-T} R F_0 \begin{pmatrix} a_1 \\ a_2 \\ a_3 \\ a_4 \end{pmatrix} = J^{-T} R F_0 \, \mathbf{a} \tag{5.18}$$

where the vector $\mathbf{f}$ contains the endpoint forces on the plane.

Now suppose you want to produce a maximal endpoint force vector in a given direction. The question is, what is the optimal activation vector $\mathbf{a}^*$ (pronounced a-star) that meets the task requirements and maximizes mechanical output?

We begin by simplifying Eq. 5.22 by defining the matrix H and its row vectors as:

$$H = J^{-T} R F_0 \tag{5.19}$$

$$H = \begin{bmatrix} \mathbf{h}_1^T \\ \mathbf{h}_2^T \end{bmatrix} \tag{5.20}$$

noting that a transposed column vector like $(\mathbf{h}_1^T)$ is, by definition, a row vector. It follows that

$$\begin{pmatrix} f_x \\ f_y \end{pmatrix} = \begin{bmatrix} \mathbf{h}_1^T \\ \mathbf{h}_2^T \end{bmatrix} \begin{pmatrix} a_1 \\ a_2 \\ a_3 \\ a_4 \end{pmatrix} \tag{5.21}$$

Phrased in this way, it is clearer that the row vectors $(\mathbf{h}_i^T)$ have physical meaning, and can be used as cost functions or linear constraint equations [4, 9, 10]. For example, suppose you want to maximize the magnitude of a force in the horizontal, positive x direction. What would be the optimal combination of muscle activations $\mathbf{a}^*$? Note that producing a force vector directed horizontally to the right means that you need to maximize the positive value of its f_x component while you minimize both the positive and negative values of its f_y component.

This problem is written in canonical form as:

$$\text{Minimize} \quad -\mathbf{h}_1^T \mathbf{a}$$
$$\text{subject to} \quad A\mathbf{a} \le \mathbf{b}$$

$$\text{where} \quad A = \begin{bmatrix} \mathbf{h}_2^T \\ -\mathbf{h}_2^T \\ \hline I_4 \\ \hline -I_4 \end{bmatrix}$$

$$\text{and} \quad \mathbf{b} = \begin{pmatrix} \varepsilon \\ \varepsilon \\ 1 \\ 1 \\ 1 \\ 1 \\ 0 \\ 0 \\ 0 \\ 0 \end{pmatrix} \tag{5.22}$$

Note several nuances:

- Given that the canonical form expects a minimization, as mentioned earlier, we needed to invert the sign of the row vector $\mathbf{h}_1^T$ so that the algorithm can maximize the f_x force component in the positive x direction.
- By convention, the elements of the activation vector $\mathbf{a}$ are bounded to be between 0 and 1, see Eq. 5.1. This is why we used the identity matrices as constraint equations.
- We want to produce 'no force output in the vertical y direction,' meaning that we want the *inequality constraints* to express an *equality constraint*. You can enforce the equality constraint

$$\mathbf{h}_2^T \mathbf{a} = f_y = b \tag{5.23}$$

where b is any scalar, as a system of two inequality constraints

$$\mathbf{h}_2^T \mathbf{a} \le b \tag{5.24}$$
$$\text{and}$$
$$\mathbf{h}_2^T \mathbf{a} \ge b \tag{5.25}$$

But it is generally not a good idea to ask a numerical algorithm to enforce strict equality constraints. Numerical and round-off errors can at times make the algorithm fail to converge because it is trying to satisfy the equality constraints to the precision of the machine representation of real numbers. In MATLAB this can be 64 bits, although it can be set by various options.

• In practice, it is generally better to limit the upper and lower bounds of f_y to a same small number (ε in this case). This is enforced as follows.

Equations 5.24 and 5.25 can be written as

$$\mathbf{h}_2^T \mathbf{a} \leq (b + \varepsilon) \tag{5.26}$$

and

$$\mathbf{h}_2^T \mathbf{a} \geq (b - \varepsilon) \tag{5.27}$$

where, using Eq. 5.15, Eq. 5.27 becomes

$$-\mathbf{h}_2^T \mathbf{a} \leq (-b + \varepsilon) \tag{5.28}$$

If $b - 0$, as in our case of having zero vertical force output, the system of inequalities becomes

$$\mathbf{h}_2^T \mathbf{a} \leq +\varepsilon \tag{5.29}$$

and

$$-\mathbf{h}_2^T \mathbf{a} \leq +\varepsilon \tag{5.30}$$

which explains the first two rows of matrix A in Eq. 5.22. ε can be in the order of 1 % or so of the smallest physically meaningful magnitude for the problem. In this case, perhaps 0.001 Newtons is adequate to represent no measurable or meaningful f_y. The result of the linear program will then satisfy that constraint to within $\pm\varepsilon$, which in practice should be equivalent to producing 'no force output in the vertical y direction.'

5.4 Geometric Interpretation of Linear Programming

Having seen the analytical formulation of linear programming, it is traditional and useful to also present it from the geometric perspective. This geometric interpretation will allow us to develop a powerful intuition about concepts like feasibility, constraints, and solution spaces that are central to understanding neuromuscular control, and many of its contemporary debates and theories.

Consider a *maximization* example where we wish to find the optimal vector $\mathbf{x} \in \mathbb{R}^2$ for the linear program

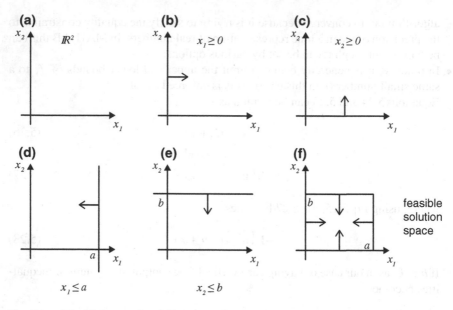

Fig. 5.1 a The 2D space of variables x_1 and x_2. **b–e** Correspondence of the first 4 inequality constrains in Eq. 5.31, to their half-planes. **f** Together they define a feasible solution space consisting of a parallelogram in the first quadrant of $\mathbb{R}^2$

$$
\begin{aligned}
\text{Maximize} \quad & x_1 + 2\,x_2 \\
\text{subject to} \quad & x_1 \geq 0 \\
& x_2 \geq 0 \\
& x_1 \leq a \\
& x_2 \leq b \\
& x_1 + x_2 \leq c
\end{aligned}
\tag{5.31}
$$

We begin by noting that the space of feasible solutions lies in $\mathbb{R}^2$, which can be visualized as the Cartesian plane. Moreover, each inequality constraint is equivalent to a linear equation that excludes a half-plane (or a half-space in higher dimensions). The constraint $x_1 \geq 0$ is equivalent to excluding the open half-plane $x_1 < 0$. Similarly, $x_1 \leq a$ excludes the open half-plane $x_1 > a$, and so on. This means that the inequality constraints that make up the matrix A, first introduced in 5.9, simply serve to reduce all of $\mathbb{R}^N$ to a subset of $\mathbb{R}^N$ where valid solutions (if any) may be found. This valid region of $\mathbb{R}^2$ can be called a *feasible solution space*. Then, the goal is to find an optimal point in that space that minimizes, or maximizes, the cost function. Figure 5.1 shows how the first 4 inequality constraints reduce the feasible solution space to a polygon (in this case a parallelogram) in the first quadrant of $\mathbb{R}^2$.

By the same logic, the last inequality constraint equation is the non-trivial line $x_1 + x_2 \leq c$ that excludes an open half-plane, and can further limit the solution space depending on the relative values of a, b, and c. Figure 5.2a, shows an example where this last inequality constraint further reduces the feasible solution space.

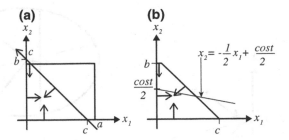

Fig. 5.2 a Example of an inequality constraint further reducing the feasible solution space. **b** Example of an instance where the value of the cost places its linear equation within the feasible solution space

Note also that the *cost function* itself is a linear equation, but whose value is yet to be determined. To see this more clearly, assume the cost is equal to the value of b. Then,

$$\begin{aligned}\text{the cost function} \quad & cost = b = x_1 + 2\,x_2 \\ \text{can be rewritten as} \quad & x_2 = -\tfrac{1}{2}\,x_1 + \tfrac{b}{2}\end{aligned} \tag{5.32}$$

the latter version being in the more commonly known slope-intercept form, with a slope of $-\tfrac{1}{2}$ and intercept of the ordinate axis at the value of $\tfrac{b}{2}$. In Fig. 5.2b we see an example where some portion of the cost function line lie within the feasible solution space. Note that in that case all points on the cost function line that also lie within the feasible solution space are valid, and produce the same cost. This is another interpretation of an underdetermined system (i.e., redundancy): There can be infinitely many valid solutions (i.e., **x** points in the feasible solution space) that produce a same result (i.e., a same cost).

Last, Fig. 5.3 shows the purpose of linear programming: Find a value of the cost (i.e., slide the cost function up and down in value) in the direction of increasing value if maximizing (or decreasing value if minimizing) until you find a unique solution for which the vector **x** is admissible, and for which the cost is at an extreme value (its maximum in this case).

Some consequences and implications of this graphical interpretation include the following:

- Linear programming can be thought of as finding the unique point where the cost function first touches the feasible region. The cost function is a straight line in $\mathbb{R}^2$, a flat plane in $\mathbb{R}^3$, and a hyper-plane in $\mathbb{R}^N$ if $N > 3$ (why?).
- A well-posed feasible region will always be a *convex set* with straight sides that meet at vertices. That is, a convex polygon in $\mathbb{R}^2$, a convex polyhedron in $\mathbb{R}^3$, and a convex polytope in $\mathbb{R}^N$ if $N > 3$ (why?).
- A unique optimal solution will always be at a vertex of the feasible region (why?).
- How would you draw a case where the problem is infeasible?

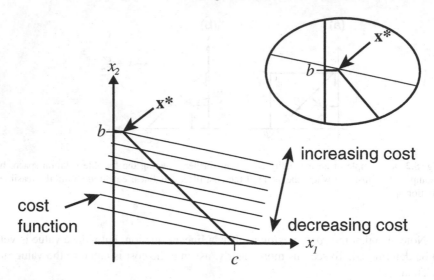

Fig. 5.3 Graphical interpretation of linear programming. Solving the linear programming problem is equivalent to finding out whether or not there exists a unique point in the feasible solution space for which the cost is greatest if maximizing, or smallest if minimizing. If it exists, that point is $\mathbf{x}^*$. Otherwise, the problem may be infeasible or unbounded, and have no such unique optimal solution

- How would you draw a case where the problem has infinitely many optimal solutions?
- How would you describe redundant constraints graphically?

Linear programming is an example of optimization in the special case where the cost function and constraints are linear. But there are entire branches of mathematics devoted to the case where the cost function and/or constraints are nonlinear, e.g., [7, 11–13]. Moreover, as described in [2], machine learning, control theory and estimation-detection theory can be used to approach the problem of neuromuscular control as one of optimal control and estimation [14, 15]. These topics will be revisited in later chapter.

5.5 Exercises and Computer Code

Exercises and computer code for this chapter in various languages can be found at http://extras.springer.com or found by searching the World Wide Web by title and author.

References

1. E.R. Kandel, J.H. Schwartz, T.M. Jessell et al., *Principles of Neural Science*, vol. 4 (McGraw-Hill, New York, 2000)
2. F.J. Valero-Cuevas, H. Hoffmann, M.U. Kurse, J.J. Kutch, E.A. Theodorou, Computational models for neuromuscular function. IEEE Rev. Biomed. Eng. **2**, 110–135 (2009)
3. F.E. Zajac, Muscle and tendon: properties, models, scaling, and application to biomechanics and motor control. Crit. Rev. Biomed. Eng. **17**(4), 359–411 (1989)
4. F.J. Valero-Cuevas, F.E. Zajac, C.G. Burgar, Large index-fingertip forces are produced by subject-independent patterns of muscle excitation. J. Biomech. **31**, 693–703 (1998)
5. V. Chvatal, *Linear Programming* (W.H. Freeman and Company, New York, 1983)
6. G. Strang, *Introduction to Linear Algebra* (Wellesley Cambridge Press, Wellesley, 2003)
7. P.E. Gill, W. Murray, M.H. Wright, *Practical Optimization* (Academic Press, New York, 1981)
8. G.B. Dantzig, *Linear Programming and Extensions* (Princeton University Press, Princeton, 1998)
9. F.J. Valero-Cuevas, Muscle coordination of the human index finger. Ph.D. thesis, Stanford University, Stanford (1997)
10. E.Y. Chao, K.N. An, Graphical interpretation of the solution to the redundant problem in biomechanics. J. Biomech. Eng. **100**, 159–167 (1978)
11. D.P. Bertsekas, *Nonlinear Programming* (Athena Scientific, 1999)
12. R. Horst, E.H. Romeijn, *Handbook of Global Optimization*, vol. 2 (Springer, Berlin, 2002)
13. D.G. Luenberger, Y. Ye, *Linear and Nonlinear Programming*. International Series in Operations Research and Management Science (2008)
14. E. Todorov, M.I. Jordan, Optimal feedback control as a theory of motor coordination. Nat. Neurosci. **5**(11), 1226–1235 (2002)
15. R. Shadmehr, S. Mussa-Ivaldi, *Biological Learning and Control: How the Brain Buildsrepresentations, Predicts Events, and Makes Decisions* (MIT Press, Cambridge, 2012)

References

1.

2.

3.

4.

5.

6.

7.

8.

9.

10.

11.

12.

13.

14.

15.

Chapter 6
The Neural Control of Musculotendon Lengths and Excursions Is Overdetermined

Abstract This chapter introduces the mathematical foundations of the concept of obligatory kinematic correlations among joint angles and musculotendon lengths. As presented in Chap. 4, tendon excursions are *overdetermined* because the angles and angle changes of the few joints uniquely determine the lengths and excursions, respectively, of *all* musculotendons. This is the opposite of redundancy: there is a single and unique set of tendon excursions that can satisfy a given limb movement. This begs the question of how the nervous system controls the excursions of all musculotendons so that the limb can move smoothly. Essentially, if for some reason any of the musculotendons undergoing an eccentric contraction fails to lengthen to satisfy the geometric requirements of the joint rotations, at the very least the limb motion will be disrupted, and at worst the limb can lock up. Physiologically, the failure to accommodate the necessary length changes could be due to anatomical interconnections among muscles or tendons, neurally mediated resistance to lengthening due to short- or long-latency *reflexes*, or spinally- and cortically-mediated commands to the muscles. This chapter lays the foundation for understanding the interactions between muscle coordination and reflex mechanisms necessary for natural movement by providing a mathematical framework for the overdetermined nature of tendon excursions. This is done for the simplified case with no anatomical interconnections among muscles or tendons, but the conclusions and intuition provided reinforce the notion that the neural control of movement for tendon-driven limbs is in fact not as redundant as is currently thought. Recall that, as mentioned in Chap. 4, the term tendon suffices for most mathematical and mechanical analyses as it applies to both robots and vertebrates. When the analysis continues on to consider muscle mechanics and its neural control, I will prefer to use the term musculotendon.

6.1 Forward and Inverse Kinematics of a Limb

While the underdetermined nature of the neural control of torque production relates to the area of limb mechanics treated in Chap. 3, the neural control of musculotendon lengths, excursions and velocities relates to limb kinematics, as seen in Chap. 2.

© Springer-Verlag London 2016
F.J. Valero-Cuevas, *Fundamentals of Neuromechanics*,
Biosystems & Biorobotics 8, DOI 10.1007/978-1-4471-6747-1_6

Therefore, we must begin with an introduction to limb kinematics. Once again, several other works present this in detail [1, 2].

As shown in Sect. 2.3, the geometric model relates joint angles to limb postures. And the Jacobian relates joint angular velocities to limb end-point velocities, Sect. 2.6. There are, of course, two ways to approach limb kinematics: by defining the joint angles and angular velocities to be the inputs that produce the endpoint location and velocities as output—the *forward kinematics* approach as defined in Sect. 2.5; or by using the *inverse kinematics* approach and treating the endpoint location and velocities as the input that defines the joint angles and angular velocities as outputs. Both approaches are equivalent in principle. But the forward kinematics problem is computationally easier because it uses the geometric model in Eq. 2.3 and the Jacobian in Eq. 2.25 directly. The inverse kinematic approach, however, needs to be done carefully because inverse trigonometric functions are not unique. Sines and cosines are periodic functions that have similar solutions with each revolution, i.e., spaced apart by multiples of 360° or 2π radians. There are many standard textbooks on kinematics to which the reader should refer, such as [1–3].

For the purpose of introducing the overdetermined nature of tendon lengths and velocities, it suffices to use the prescribed or measured joint angles and angular velocities for a given task. Knowing these joint angles and their sequences, we can find the tendon excursions and velocities on the basis of the moment arm matrix introduced in Eq. 4.26.

6.2 Forward Kinematics of a 5 DOF Arm

Let us use the forward kinematic approach to find the tendon excursions and velocities for a human arm model assumed to have 5 DOFs, Fig. 6.1. In principle we only need to know the moment arm matrix of the limb, but for completeness I will present the full forward kinematic model to place the results in a functional context. This kinematic model of the arm is inspired by the anatomical DOFs in [4] and published in [5].

Using Eqs. 2.7 and 2.10 we can define the sequential rotational DOFs, which follow the general case of configuration coordinate transformation matrices presented in Eq. 2.4:

1. Shoulder external rotation, about the collinear i_0 and i_1 axes

$$T_0^1 = \begin{bmatrix} 1 & 0 & 0 & 0 \\ 0 & c_1 & -s_1 & 0 \\ 0 & s_1 & c_1 & 0 \\ 0 & 0 & 0 & 1 \end{bmatrix} \tag{6.1}$$

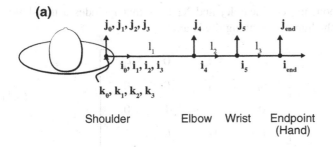

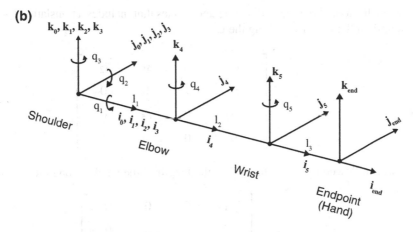

Fig. 6.1 A 3-link, 5 DOF simplified human arm model. Its fixed base is the shoulder and its endpoint is the hand as a whole. **a** Top view. **b** 3D view

2. Shoulder adduction, about the collinear j_1 and j_2 axes

$$T_1^2 = \begin{bmatrix} c_2 & 0 & s_2 & 0 \\ 0 & 1 & 0 & 0 \\ -s_2 & 0 & c_2 & 0 \\ 0 & 0 & 0 & 1 \end{bmatrix} \tag{6.2}$$

3. Shoulder horizontal adduction, about the collinear k_2 and k_3 axes

$$T_2^3 = \begin{bmatrix} c_3 & -s_3 & 0 & 0 \\ s_3 & c_3 & 0 & 0 \\ 0 & 0 & 1 & 0 \\ 0 & 0 & 0 & 1 \end{bmatrix} \tag{6.3}$$

4. Elbow flexion, about the collinear $\mathbf{k}_3$ and $\mathbf{k}_4$ axes that includes a translation l_1—the length of the humerus—along the $\mathbf{i}_3$ axis

$$T_3^4 = \begin{bmatrix} c_4 & -s_4 & 0 & l_1 \\ s_4 & c_4 & 0 & 0 \\ 0 & 0 & 1 & 0 \\ 0 & 0 & 0 & 1 \end{bmatrix} \tag{6.4}$$

5. Wrist flexion, about the collinear $\mathbf{j}_4$ and $\mathbf{j}_5$ axes that includes a translation l_2—the length of the forearm—along the $\mathbf{i}_4$ axis

$$T_4^5 = \begin{bmatrix} c_5 & 0 & s_5 & l_2 \\ 0 & 1 & 0 & 0 \\ -s_5 & 0 & c_5 & 0 \\ 0 & 0 & 0 & 1 \end{bmatrix} \tag{6.5}$$

and is followed by a translation l_3—the length of the hand—along the $\mathbf{i}_5$ axis

$$T_5^{end} = \begin{bmatrix} 1 & 0 & 0 & l_3 \\ 0 & 1 & 0 & 0 \\ 0 & 0 & 1 & 0 \\ 0 & 0 & 0 & 1 \end{bmatrix} \tag{6.6}$$

This produces the complete transformation that describes the location and orientation of the endpoint of the hand (frame 5) in the coordinates of the shoulder (frame 0), as shown in Eq. 6.7.

$$\begin{aligned} \mathbf{v}_0 &= T_0^1 \; T_1^2 \; T_2^3 \; T_3^4 \; T_4^5 \; T_5^{end} \; \mathbf{v}_{end} \\ T_0^{end} &= T_0^1 \; T_1^2 \; T_2^3 \; T_3^4 \; T_4^5 \; T_5^{end} \\ \mathbf{v}_0 &= T_0^{end} \; \mathbf{v}_{end} \end{aligned} \tag{6.7}$$

These coordinate transformation matrices are long trigonometric expressions that are left as exercises to the reader. Suffice it to say that from them you can obtain the geometric model as shown in Sect. 3.2. This is the forward kinematic problem because the inputs are the joint angles $\mathbf{q}$, and the outputs are the location and orientation of the endpoint. Section 3.2 also shows how the Jacobian is obtained from the geometric model.

6.3 Inverse Kinematics of a Limb

In this case, we know or want a particular location and orientation of the endpoint
in space, and we want to find the joint angles that achieve it. In principle it simply
requires one to invert the geometric model. This nonlinear function is amenable to
any number of analytical and numerical approaches to invert a system of nonlinear
equations [3]. More generally, you want to find the *sequence* of joint angles that
produces a given *trajectory* of the endpoint. Here I will consider three methods, but
the reader can explore the many other alternatives relevant to their particular goal
and model.

6.3.1 Closed Form Analytical Approach

Consider the case where the geometric model leads to an invertible Jacobian, where
you have as many joint angles as elements of the output wrench as described in
Sect. 3.5. In this case the inverse kinematic problem can be phased as follows: Find a
posture of the limb for which you know the location of the endpoint is on a particular
trajectory of interest, preferably its start. This trajectory is of course already known
to you, such as that of reaching from one point to another. This posture should be
valid in the sense that the joint angles should be achievable by the robot or vertebrate
limb. Then take steps away from that point along the trajectory to build a time history
of joint angles that produce that trajectory.

Figure 6.2 shows the limb at the starting posture that produces the endpoint location (x_0, y_0); call this q_0. At that posture the Jacobian of the limb is $J(q_0)$. Given
that Eq. 2.26, which is

$$\dot{q} = J(q)^{-1}\dot{x} \tag{6.8}$$

Fig. 6.2 Schematic
description of inverse
kinematics using a closed
form analytical approach

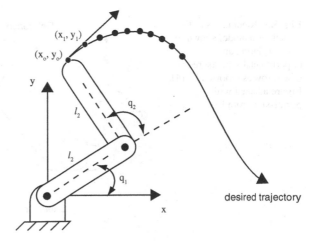

can be expressed in discrete form as

$$\mathbf{\Delta q} = J(\mathbf{q})^{-1} \, \mathbf{\Delta x} \tag{6.9}$$

we can use the following equation to bootstrap across discrete postures from $i = 0$ to the end of the discretized trajectory.

$$\mathbf{\Delta q}_{i+1} = J(\mathbf{q}_i)^{-1} \, \mathbf{\Delta x}_i \tag{6.10}$$

Note that this is not unlike an Euler integration method, or an initial value problem, where you know the gradient of a function at any point, and you use that gradient plus an initial condition to take steps forward in time or space.

6.3.2 Numerical Approach

Alternatively, you can use an existing nonlinear numerical optimization algorithm such as lsqnonlin in MATLAB to solve the system of equations. But note that you must provide appropriate constraints on joint angles so as not to get unrealistic mirror-image postures of the limb (i.e., with the knee or elbow hyper-extended).

6.3.3 Experimental Approach

Last, you can extract the joint angles from experimental measurements. Video, motion capture, inertial measurement units, etc. are used routinely to measure limb movements by locating fiducial markers on the body—and then either human operators or algorithms are used to extract the joint angles. These classical methods

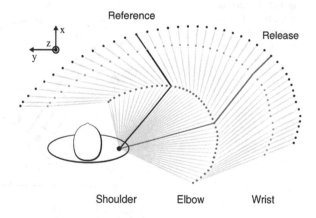

Fig. 6.3 Kinematics of the 5 DOF arm model shown in Fig. 6.1 during an experimentally measured disc throw as reported by [4]. Figure adapted with permission from [5]

commonly use regression models, model-based estimation, Kalman filters, etc., and are described in, say, [6, 7]. More recently consumer products have been developed for 'markerless' motion capture [8].

As an example, Fig. 6.3 takes the joint angles extracted from motion capture recordings on an arm during a disc throw from [4], and plots them using the 5 DOF arm model in Fig. 6.1 and Eq. 6.7.

6.4 The Overdetermined Problem of Tendon Excursions

Now that we have the basic kinematic concepts in place, we can define the problem of overdetermined tendon excursions. First, we need some definitions that go beyond those presented in Chap. 4. *Tendon excursion* is a clinical term that relates the distance a tendon traverses as the limbs move and muscles contract. Mechanically speaking, we need to be more precise to understand the overall changes in the length of the musculotendon. As discussed in Chap. 4, musculotendons are the combined entity of the tendon of origin, the muscle fibers, and the tendon of insertion [9].

6.4.1 Tendon Excursion Versus Musculotendon Excursion

In vertebrates, the viscoelastic properties of the musculotendon matter [9, 11]. The change in musculotendon length is distributed among its tendinous and muscular elements. In some cases, tendons tend to be stiffer than muscle tissue. The series arrangement of muscle fibers and tendons can in some cases allow the assumption that tendon excursions mostly affect the length of the muscle fibers.

But such simplifications need to be made carefully. Consider the case of the static condition where a person produces a static force with their limb, but the limb does not change posture; also called an isometric (same-length) contraction, like when you push against a wall. Even though there are no changes in joint angles, the elasticity of the tendons will allow them to stretch while the muscles shorten—slightly perhaps, but shorten nonetheless. Sometimes muscle fibers can even be shortening while the overall length of the musculotendon is increasing [12–14]. That is, paradoxical changes in muscle fiber length can occur due to the interplay of inertial and muscle forces, muscle activation, joint rotations, tendon excursions, and viscoelastic properties of tissue. Therefore, tendon excursions do not necessarily define muscle fiber lengths or velocities. For this one needs to consider tendon compliance and stretch.

Considering that joint rotations and tendon excursions occur in a given amount of time, tendon excursions can allow us—if we assume stiff tendons or consider their properties—to calculate *musculotendon length* and *musculotendon velocity*, and thus *muscle fiber length* and *muscle fiber velocity*, as they are called in the muscle mechanics literature [9]. For the sake of simplicity, and without loss of generality, we can assume that tendons are stiff enough such that Eq. 4.27, which is

$$\delta s = -R(\mathbf{q})^T \, \delta \mathbf{q} \tag{6.11}$$

can be re-written in terms of muscle fiber length, $\mathbf{l}_m$, and muscle fiber velocity, $\mathbf{v}_m$ as

$$\delta \mathbf{l}_m = -R(\mathbf{q})^T \, \delta \mathbf{q} \tag{6.12}$$

$$\mathbf{v}_m = -R(\mathbf{q})^T \, \dot{\mathbf{q}} \tag{6.13}$$

This formulation is revealing from a variety of perspectives. A direct geometric consequence of Eqs. 6.12 and 6.13 is that the DOFs $\mathbf{q}$ determine the length and velocity of *all* muscles. As mentioned in Eq. 4.28, $R(\mathbf{q})^T \in \mathbb{R}^{N \times M}$ where $M < N$.

Thus the mapping from joint angles to tendon excursions is from a lower to a higher dimension. This is the opposite of redundancy, as tendon excursions are uniquely defined by the geometry of the limb and the characteristics of the movement—and the nervous system has little say in the matter. All muscles that must lengthen to accommodate limb motion must be allowed by the nervous system to attain those necessary muscle fiber lengths and velocities.

Recall that the nervous system only has direct command over the activation and reflex loops of a given muscle—but whether or not a particular muscle shortens (*concentric contraction*), stays the same length (*isometric contraction*), or lengthens (*eccentric contraction*) depends entirely on its short- and long-latency reflexes, the mechanical state and forces acting on the limb, and the activation of all other muscles involved. Thus, if the lengthening of a given muscle is not forthcoming, the required postures will not be attainable or the movement will be disrupted. For a single joint movement, the DOF can cease if even one muscle that needs to lengthen fails to do so. Of the muscles that need to shorten to accomplish the motion, those that are not controlled properly will simply go slack. But at least some of those muscles need to produce the necessary forces to move the limb.

6.4.2 Muscle Mechanics

A muscle's ability to produce a force depends critically on the length and velocity of its fibers. Figure 6.4 describes nominal force–length and force–velocity properties as per current thinking [9, 15, 16]. Given that muscle fibers have a length and a velocity at any given point in time, these properties are best visualized as a 3D surface that contains all combinations of muscle fiber lengths and muscle fiber velocities, Fig. 6.5. However, despite decades of research, it is not known if these force–length and force–velocity properties in fact combine and superimpose in such a linear way. These properties need to be treated with caution, and some skepticism [5, 17–22], and remain active fields of research [16, 18, 23–26]. But force–length and force–velocity

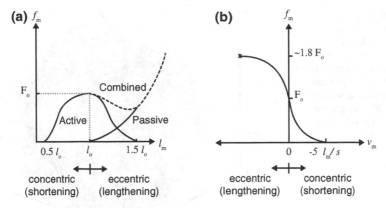

Fig. 6.4 Nominal length and velocity properties of maximally activated muscle tissue. These properties are often described in normalized muscle fiber length and velocity, where the maximal *isometric* (i.e., static and constant length) force a muscle can produce is F_0. Note how the directions of contraction are traditionally indicated in the abscissa of the figures, which define *concentric* and *eccentric* contractions. **a** Isometric force–length (f–l) properties of a tendon-less muscle when fully activated. The passive force is generated by the muscle's connective tissue when stretched. Active forces are produce by the sarcomeres, the molecular motors that are the foundation of neurally-controlled muscle activity [15, 16]. The active force is assumed to scale linearly with activation, whereas the passive force is assumed to be independent of activation. By definition, peak active isometric force, F_0, is developed when fibers are at their optimal length (i.e., when $l_m = l_0$). **b** Force–velocity (f–v) relation of maximally activated muscle tissue. The force at zero normalized fiber velocity is F_0. An applied constant force less (greater) than F_0 causes muscle tissue to shorten (lengthen). Limits exist to force generation ($<1.8\,F_0$) and active shortening velocity ($v_{max} = -5\frac{l_m}{s}$, maximal shortening velocity). These relationships are assumed to scale linearly for less than fully-activated muscle tissue [9, 15]. Adding a tendon will affect these properties due to its compliance [9]

effects are undeniable, and the available numerical approximations to them have nevertheless provided good working hypotheses to create and use musculoskeletal models [27].

As shown in Figs. 6.4 and 6.5, the capacity to produce static force suffers greatly if muscle fibers are either too short or too long. For a detailed treatment of this, please see [16]. Briefly, during a concentric contraction the force capacity drops precipitously as the shortening velocity increases. To the point that muscles are thought not to be able to produce any concentric forces beyond v_{max}. This maximal velocity is considered to be about 5 muscle fiber lengths per second ($v_{max} = -5\frac{l_m}{s}$). Conversely, because muscles can only actively pull and not push, an eccentric contraction is by definition a resistance to an externally imposed lengthening. In this case, muscles can produce a resistive force greater than their static maximal force F_0 by about 50 % because the crossbridges in their sarcomeres are being pulled apart, which takes more force than the force the crossbridges can produce when under isometric conditions. However, such forced stretching of a muscle can and does cause injury, and at high enough velocities it can injure the muscle to the point of rupture.

Fig. 6.5 Linear
superimposition of
force–length and
force–velocity properties
of a tendon-less muscle. The
3D surface interpolates
across combinations of
muscle fiber length and
muscle fiber velocities.
Courtesy of
E. Todorov

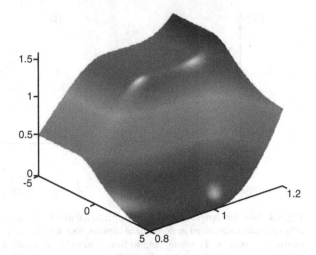

It is important to mention that by controlling and coordinating concentric versus eccentric contractions, the nervous system is in fact controlling the mechanical work that muscles perform. Muscles produce positive work (i.e., inject kinetic and potential energy into the limbs and environment) in concentric contractions because muscle force and tendon excursion are in the same direction, and produce negative work (i.e., operate as breaks, or energy sinks, removing kinetic and potential energy from the limbs and environment) in eccentric contractions as the force and displacement are in opposite directions [28].

When calculating mechanical work, it is important to know how external work is related to the definition of positive and negative tendon excursion in Sect. 4.6. Active shortening of the musculotendon (i.e., *negative* tendon excursion during a concentric contraction, Fig. 6.4b) while under tension means the muscle is producing *positive* work. This is because the force in the tendon is in the same direction as the movement of the tendon.

Conversely, a *positive* tendon excursion is one that lengthens the musculotendon,[1] Fig. 6.4b. In such eccentric contractions, the muscle serves as a break that absorbs energy (i.e., produces *negative* work as the force and displacement are in opposite directions). Understanding how the nervous system regulates the exchange of mechanical energy among the body, limbs, and environment is central to our understanding of all behavior and the reader is referred to that body of work (e.g., [5, 28–31], and references therein).

---

[1]Recall that tendon excursions do not obligatorily define muscle fiber length changes. The latter depend on the interaction among the viscoelastic properties of tendon and muscle, as mentioned above. Only the length change of the muscle fibers determines eccentric versus concentric contractions in muscle. For the sake of simplicity and without loss of generality, I assume stiff tendons. But the correction for this assumption can be added depending on the goal of the model.

6.4.3 Reflex Mechanisms Interact with Limb Kinematics, Mechanics, and Muscle Properties

An additional topic to be considered before working on the overdetermined nature of the neural control of muscle excursions is that of reflex mechanisms. By now you must be considering the objection that studying tendon excursions assumes that muscles are always taut. That is, that muscles are not slack at any time. If a muscle is slack, then limb movements do not depend in it, it cannot disrupt the movement, nor suffer injury from eccentric contractions.

However, healthy muscles are never slack by virtue of *reflex* mechanisms that produce muscle tone. In fact, a muscle without tone is considered pathologic, as it is either paralyzed or unnaturally flaccid. For a detailed discussion of spinal reflex mechanisms such as the monosynaptic stretch reflex see [32]. Briefly, the *monosynaptic stretch reflex* is a rapid, automatic, and involuntary resistance to musculotendon lengthening. Reflexes can be either spinal or transcortical, monosynaptic or olygosynaptic, impede or facilitate stretch, etc. [32]. In the particular case of monosynaptic stretch reflexes, an eccentric contraction that is fast enough will elicit an initial resistance to stretch within a latency of around 30–50 ms. Some pathological conditions have hyperactive stretch reflexes of spinal or cortical origin that, in fact, lead to spastic, erratic and otherwise irregular motions. See for example [33], and the references therein.

The idea that coordinating, regulating, and mitigating reflex responses is critical to natural movement was first introduced by Sir Charles Sherrington over a hundred years ago [34]. It is one of the oldest and most central tenets of sensorimotor function. However, it remains poorly understood despite decades of work [35, 36], and is seldom considered in computational approaches to neuromuscular control of limbs [12]. The details of how the nervous system coordinates muscle activations to produce smooth and rapid limb motion when healthy muscles naturally resist being stretched remains unanswered for multi-joint, multi-muscle limbs, and is an area of active research [37–39]. This is critical to, for example, understanding disability following stroke [36] or cerebral palsy [33], and the design of rehabilitation strategies with externally imposed limb motions (e.g., [40]).

6.5 Example of a Disc Throw Motion with a 17-Muscle, 5 DOF Arm

It is now clear how rapid and smooth limb movements—such as those necessary for everyday life and sports—require the nervous system to control all muscles in a way that allows well-coordinated concentric and eccentric contractions, at velocities that do not extinguish the ability to produce muscle force, or induce damage, while also keeping reflex responses at bay. These multiple, obligatory, and simultaneous constraints that confront the nervous system are the opposite of redundancy: neural control is very limited in the coordination strategies it can feasibly use [5, 17].

Table 6.1 Anatomical details of 17-arm muscles. DOF labels correspond to those in Fig. 6.1b

Muscle	Symbol	DOF	Action	Optimal length (cm)	Moment arm (cm)	Pennation angle (°)
Latissimus dorsi	M1	q1, q2, q3	Int/Ext Rot Abd/Add Horiz. Abd/Add	39.27	3.3	21.6
Pectoralis major	M2	q1, q2, q3	Int/Ext Rot Abd/Add Horiz. Abd/Add	19	1.9	22.33
Teres major	M3	q1, q2	Int/Ext Rot Abd/Add	14.84	2	16
Teres minor	M4	q1, q3	Int/Ext Rot Horiz. Abd/Add	5.72	1.8	24
Deltoid (Anterior)	M5a	q1, q2, q3	Int/Ext Rot Abd/Add Horiz. Abd/Add	12.8	1.9	22
Deltoid (Posterior)	M5b	q1, q2, q3	Int/Ext Rot Abd/Add Horiz. Abd/Add	12.8	0.8	18
Subscapularis	M6	q1	Int/Ext Rot	8.92	1.9	20
Supraspinatus	M7	q2	Abd/Add	4.28	1	7
Infraspinatus	M8	q1, q3	Int/Ext Rot Horiz. Abd/Add	6.76	2.3	19
Coracobrachilis	M9	q2, q3	Abd/Add Horiz.Abd/Add	17.6	2	27
Biceps brachii	M10	q2, q4	Flex/Ext	14.22	3.6	0
Brachialis	M11	q4	Flex/Ext	10.28	1.8	0
Triceps brachii	M12	q2, q4	Flex/Ext	8.77	2.1	10
Flexor carpi radialis	M13	q5	Flex/Ext	5.1	1.4	3
Flexor carpi ulnaris	M14	q5	Flex/Ext	3.98	1.9	12
Extensor carpi radialis longus	M15	q5	Flex/Ext	7.28	1.8	0
Extensor carpi ulnaris	M16	q5	Flex/Ext	3.56	2.3	4

These data can be used to assemble a moment arm matrix for the 5 DOF, 17-muscle arm model. These moment arm values are compiled from [5, 41, 42]

A relatively realistic model of a human arm could include 17 independent muscles. As in [5], these muscles can be those listed in Table 6.1. From such a table of anatomical data one can assemble a moment arm matrix that, as per the convention in Eq. 4.21, has as many rows as there are degrees of freedom (i.e., 5), and as many columns as there are muscles (i.e., 17). Such a moment arm matrix can be transposed to calculate tendon excursions as per Eq. 4.26.

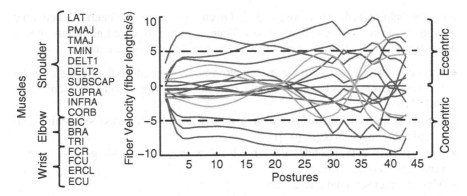

Fig. 6.6 Normalized muscle fiber velocities during the disc throw motion shown in Fig. 6.3 for an assumed duration of 450 ms. Muscle fiber velocities in many muscles throughout the motion are either too fast concentrically (< -5 muscle fiber lengths per second), or eccentrically ($> +5$ muscle fiber lengths per second). Thus they should not be able to produce produce force and contribute to the throw, or be at a high risk of rupture, respectively. For additional details see [5]

Consider the disc throw motion shown in Fig. 6.3 using the moment arm matrix for the arm as per the data in Table. 6.1, the kinematic model in Fig. 6.1, and the time-histories of joint angles from [4]. From this we can calculate the muscle velocities for all 17 muscles in the model. We assumed, conservatively, that the initiation of forward motion, release, and follow-through portions of the throw lasted 450 ms; and approximated it as 45 unique postures at 10 ms time steps, as illustrated in Fig. 6.3. Figure 6.6 shows that the muscle fiber velocities in many muscles are either very fast concentrically (< -5 muscle fiber lengths per second), or eccentrically (< 5 muscle fiber lengths per second) throughout the motion. So these muscles are either close to losing their ability to produce concentric force, or are at risk of rupturing eccentrically, respectively. For an extended discussion of these results, see [5].

6.6 Implications to Neural Control and Muscle Redundancy

This overdetermined problem raises several important questions and suggests several lines of research. Given that the kinematics of the arm are objectively correct, and that Eqs. 6.12 and 6.13 are geometrically given, the logical conclusion is that our understanding of neural control, muscle redundancy, muscle mechanics, and reflex mechanisms needs to be revised and expanded to allow for these mechanically-driven robust results as reported in [5].

More specifically, what inhibits reflex mediated responses to allow such coordinated obligatory concentric-eccentric contractions? α-γ co-activation, reciprocal inhibition, cortically-mediated presynaptic inhibition, and gating of spindle afferent information are some of neural interactions thought to be necessary to modulate/inhibit reflex responses [32]. Given that these are neurally-mediated responses, the nervous system must issue neural commands, coordinated throughout the entire duration of the movement to, at the very least, simultaneously:

- Regulate activity across the α-motoneuron pools to produce the necessary joint torques as per the classical muscle redundancy force sharing problem (e.g., [43, 44])
- Coordinate reciprocal inhibition of α-motoneuron pools across shortening and lengthening muscles (e.g., [45])
- Tune γ-drive and/or inhibit the stretch reflex in muscles undergoing eccentric contractions (e.g., [46])
- Mediate interneuronal interactions
- Satisfy the temporal constraints of conduction velocities, muscle excitation-contraction dynamics, and activation/deactivation time constants [9] to ensure the continuity of these neural commands as the motion progresses

This compounding of multiple spatial and temporal constraints naturally leads to a shrinking of the set of feasible neural commands for natural movements (see Introduction and [17, 47]). In fact, clinicians have long been aware of how disorders of reflexes or the neural circuits of 'afferented muscles' lead to disruptions or failures of movements. For an overview see [33, 48]. Thus, these movement pathologies may in fact be a natural consequence of the nervous system failing to meet the stringent spatio-temporal demands listed above.

This again supports the view that extant vertebrates must solve a control problem that goes beyond the force-sharing problem as per the classical notion of muscle redundancy; and thus neural control of behavior is likely not as redundant as currently thought. In Part III, I present a mathematically rigorous definition to the notion that there is such a thing as a *set of feasible neural commands* at any point in time, and use it to develop an approach to understanding the spatio-temporal nature of the neural control of tendon-driven limbs.

Last, this analysis made the reasonable assumption that tendons are stiff enough for Eqs. 6.12 and 6.13 to apply. However, overdetermined systems of equations can have at most one solution, or typically no solution at all. As can be seen from the list above, the system is overdetermined in space (i.e., geometrically) and time (i.e., sequencing, muscle excitation-contraction dynamics, activation/deactivation time constants, etc.). As mentioned in Sect. 4.6, practical—albeit inexact—solutions to overdetermined systems are found by finding solutions that violate the constraints as little as possible, such as a those found by least squares error methods. In this case, the fact that muscles and tendons have non-trivial amounts of passive elasticity may in fact be the biological analog to a means to allow a least squares error solution. Such passive elasticity would provide tolerance to errors as the neural control fails to satisfy all spatio-temporal constraints on tendon excursions or velocities. From the engineering perspective, such elasticity in the actuators is considered problematic because it complicates the control of tendons by adding delays and internal actuator dynamics, and reduces actuator bandwidth. But in the case of anatomical tendon-driven limbs, this built-in tolerance to excursion errors may be a critical complement to, and enabler of, the neural control of smooth movements [5].

In addition, several authors have highlighted the tradeoffs for stiff versus compliant tendons in terms of neural control versus energy dissipation and regulation. For example, Zajac [9, 49] has highlighted the fact that the elasticity of the tendon modifies the force–length properties of muscle—which are usually given for tendon-less muscles as in Fig. 6.4. Others have highlighted and debated the importance of tendons to locomotor function from the mechanical, shock-absorption, and energetic perspectives (e.g. [50–54]). This continues to be an active area of research in neural control, muscle mechanics, biomechanics, and evolutionary biology.

6.7 Exercises and Computer Code

Exercises and computer code for this chapter in various languages can be found at http://cxtras.springer.com or found by searching the World Wide Web by title and author.

References

1. R.M. Murray, Z. Li, S.S. Sastry, *A Mathematical Introduction to Robotic Manipulation* (CRC Press, 1994)
2. T. Yoshikawa, *Foundations of Robotics: Analysis and Control* (MIT Press, Cambridge, 1990)
3. O. Bottema, B. Roth, *Theoretical Kinematics* (Dover Publications, 2012)
4. S.A. Hummel, Frisbee flight simulation and throw biomechanics. Master's thesis, University of California, (2003)
5. F.J. Valero-Cuevas, B.A. Cohn, H.F. Yngvason, E.L. Lawrence, Exploring the high-dimensional structure of muscle redundancy via subject-specific and generic musculoskeletal models. J. Biomech. **48**(11), 2887–2896 (2015)
6. D.A. Winter, *Biomechanics and Motor Control of Human Movement* (Wiley, NewYork, 2009)
7. E. Todorov, Probabilistic inference of multijoint movements, skeletal parameters and marker attachments from diverse motion capture data. IEEE Trans. Biomed. Eng. **54**(11), 1927–1939 (2007)
8. J. Shotton, T. Sharp, A. Kipman, A. Fitzgibbon, M. Finocchio, A. Blake, M. Cook, R. Moore, Real-time human pose recognition in parts from single depth images. Commun. ACM **56**(1), 116–124 (2013)
9. F.E. Zajac, Muscle and tendon: properties, models, scaling, and application to biomechanics and motor control. Crit. Rev. Biomed. Eng. **17**(4), 359–411 (1989)
10. Y.T. Lee, H.R. Choi, W.K. Chung, Y. Youm, Stiffness control of a coupled tendon-driven robot hand. IEEE Control Syst. **14**(5), 10–19 (1994)
11. R. Alexander, G.M.O. Maloiy, R.F. Ker, A.S. Jayes, C.N. Warui, The role of tendon elasticity in the locomotion of the camel (camelus dromedarius). J. Zool. **198**(3), 293–313 (1982)
12. L.A. Elias, R.N. Watanabe, A.F. Kohn, Spinal mechanisms may provide a combination of intermittent and continuous control of human posture: predictions from a biologically based neuromusculoskeletal model. PLoS Comput. Biol. **10**(11), e1003944 (2014)

13. I.D. Loram, C.N. Maganaris, M. Lakie, Active, non-spring-like muscle movements in human postural sway: how might paradoxical changes in muscle length be produced? J. Physiol. **564**(1), 281–293 (2005)
14. R.I. Griffiths, Shortening of muscle fibres during stretch of the active cat medial gastrocnemius muscle: the role of tendon compliance. J. Physiol. **436**(1), 219–236 (1991)
15. T.A. McMahon, *Muscles, Reflexes, and Locomotion* (Princeton University Press, New Jersey, 1984)
16. R.L. Lieber, *Skeletal Muscle Structure, Function, and Plasticity* (Lippincott Williams & Wilkins, Philadelphia, 2002)
17. K.G. Keenan, V.J. Santos, M. Venkadesan, F.J. Valero-Cuevas, Maximal voluntary fingertip force production is not limited by movement speed in combined motion and force tasks. J. Neurosci. **29**, 8784–8789 (2009)
18. A.A. Biewener, J.M. Wakeling, S.S. Lee, A.S. Arnold, Validation of hill-type muscle models in relation to neuromuscular recruitment and force–velocity properties: predicting patterns of in vivo muscle force. Integr. Comp. Biol. **54**(6), 1072–1083, page icu070, (2014)
19. R.H. Miller, A comparison of muscle energy models for simulating human walking in three dimensions. J. Biomech. **47**(6), 1373–1381 (2014)
20. G.A. Tsianos, C. Rustin, G.E. Loeb, Mammalian muscle model for predicting force and energetics during physiological behaviors. IEEE Trans. Neural Syst. Rehabil. Eng. **20**(2), 117–133 (2012)
21. M.R. Rehorn, A.K. Schroer, S.S. Blemker, The passive properties of muscle fibers are velocity dependent. J. Biomech. **47**(3), 687–693 (2014)
22. E.J. Cheng, I.E. Brown, G.E. Loeb, Virtual muscle: a computational approach to understanding the effects of muscle properties on motor control. J. Neurosci. Methods **101**(2), 117–130 (2000)
23. K.C. Nishikawa, J.A. Monroy, T.E. Uyeno, S.H. Yeo, D.K. Pai, S.L. Lindstedt, Is titin a 'winding filament'? A new twist on muscle contraction. Proc. R. Soc. Lond. B: Biol. Sci. **279**(1730), 981–990 (2012)
24. W. Herzog, M. Duvall, T.R. Leonard, Molecular mechanisms of muscle force regulation: a role for titin? Exerc. Sport. Sci. Rev. **40**(1), 50–57 (2012)
25. T.R. Leonard, W. Herzog, Regulation of muscle force in the absence of actin-myosin-based cross-bridge interaction. Am. J. Physiol.-Cell Physiol. **299**(1), C14–C20 (2010)
26. S. Dayanidhi, R.L. Lieber, Skeletal muscle satellite cells: mediators of muscle growth during development and implications for developmental disorders. Muscle Nerve **50**(5), 723–732 (2014)
27. F.J. Valero-Cuevas, H. Hoffmann, M.U. Kurse, J.J. Kutch, E.A. Theodorou, Computational models for neuromuscular function. IEEE Rev. Biomed. Eng. **2**, 110–135 (2009)
28. B.W. Tobalske, T.L. Hedrick, K.P. Dial, A.A. Biewener, Comparative power curves in bird flight. Nature **421**(6921), 363–366 (2003)
29. A.A. Biewener, Patterns of mechanical energy change in tetrapod gait: pendula, springs and work. J. Exp. Zool. Part A: Comp. Exp. Biol. **305**(11), 899–911 (2006)
30. J.M. Donelan, R. Kram, A.D. Kuo, Simultaneous positive and negative external mechanical work in human walking. J. Biomech. **35**(1), 117–124 (2002)
31. M.J. Srinivasan, A. Ruina, Computer optimization of a minimal biped model discovers walking and running. Nature **439**(7072), 72–75 (2006)
32. E. Pierrot-Deseilligny, D. Burke, *The Circuitry of the Human Spinal Cord: Its Role in Motor Control and Movement Disorders* (Cambridge University Press, Cambridge, 2005)
33. T.D. Sanger, D. Chen, D.L. Fehlings, M. Hallett, A.E. Lang, J.W. Mink, H.S. Singer, K. Alter, H. Ben-Pazi, E.E. Butler et al., Definition and classification of hyperkinetic movements in childhood. Mov. Disord. **25**(11), 1538–1549 (2010)
34. C.S. Sherrington, Reflex inhibition as a factor in the co-ordination of movements and postures. Exp. Physiol. **6**(3), 251–310 (1913)
35. C. Capaday, R.B. Stein, Amplitude modulation of the soleus H-reflex in the human during walking and standing. J. Neurosci. **6**(5), 1308–1313 (1986)

36. X. Hu, N.L. Suresh, W.Z. Rymer, Estimating the time course of population excitatory postsynaptic potentials in motoneurons of spastic stroke survivors. J. Neurophysiol. **113** (6), 1952–1957, pp. jn-00946, (2014)
37. C.M. Niu, S.K. Nandyala, T.D. Sanger, Emulated muscle spindle and spiking afferents validates vlsi neuromorphic hardware as a testbed for sensorimotor function and disease. Front. Comput. Neurosci. **8**, (2014)
38. C.M. Niu, J. Rocamora, W.J. Sohn, F.J. Valero-Cuevas, T.D. Sanger, Force–velocity property of muscle is critical for stabilizing a tendon-driven robotic joint controlled by neuromorphic hardware, in *Proceedings of the 6th International IEEE/EMBS Conference of Neural Engineering.* (IEEE/EMBS, 2013)
39. T. Buhrmann, E.A. Di Paolo, Spinal circuits can accommodate interaction torques during multijoint limb movements. Front. Comput. Neurosci. **8**, (2014)
40. Y. Masugi, T. Kitamura, K. Kamibayashi, T. Ogawa, T. Ogata, N. Kawashima, K. Nakazawa, Velocity-dependent suppression of the soleus H-reflex during robot-assisted passive stepping. Neurosci. Lett. **584**, 337–341 (2015)
41. B.A. Garner, M.G. Pandy, Estimation of musculotendon properties in the human upper limb. Ann. Biomed. Eng. **31**(2), 207–220 (2003)
42. R.V. Gonzalez, T.S. Buchanan, S.L. Delp, How muscle architecture and moment arms affect wrist flexion-extension moments. J. Biomech. **30**(7), 705–712 (1997)
43. E.Y. Chao, K.N. An, Graphical interpretation of the solution to the redundant problem in biomechanics. J. Biomech. Eng. **100**, 159–167 (1978)
44. B.I. Prilutsky, Muscle coordination: the discussion continues. Mot. Control **4**(1), 97–116 (2000)
45. H. Hultborn, Spinal reflexes, mechanisms and concepts: from Eccles to Lundberg and beyond. Prog. Neurobiol. **78**(3), 215–232 (2006)
46. E.P. Zehr, R.B. Stein, What functions do reflexes serve during human locomotion? Prog. Neurobiol. **58**(2), 185–205 (1999)
47. G.E. Loeb, Overcomplete musculature or underspecified tasks? Mot. Control **4**(1), 81–83 (2000)
48. T.D. Sanger, D. Chen, M.R. Delgado, D. Gaebler-Spira, M. Hallett, J.W. Mink et al., Definition and classification of negative motor signs in childhood. Pediatrics **118**(5), 2159–2167 (2006)
49. F.E. Zajac, How musculotendon architecture and joint geometry affect the capacity of muscles to move and exert force on objects: a review with application to arm and forearm tendon transfer design. J. Am. Hand Surg. **17**(5), 799–804 (1992)
50. A. Andrew, Muscle and tendon contributions to force, work, and elastic energy savings: a comparative perspective. Exerc. Sport. Sci. Rev. **28**(3), 99–107 (2000)
51. N. Konow, T.J. Roberts, The series elastic shock absorber: tendon elasticity modulates energy dissipation by muscle during burst deceleration. Proc. R. Soc. Lond. B: Biol. Sci. **282**(1804):2014–2800 (2015)
52. C. Hanson-Carbonneau, M. Eng, A.S. Arnold, D.E. Lieberman, A.A. Biewener, The capacity of the human iliotibial band to store elastic energy during running. J. Biomech. (2015)
53. R.M. Alexander, Tendon elasticity and muscle function. Comp. Biochem. Physiol. Part A: Mol. Integr. Physiol. **133**(4), 1001–1011 (2002)
54. C.N. Holt, T.J. Roberts, G.N. Askew, The energetic benefits of tendon springs in running: is the reduction of muscle work important? J. Exp. Biol. **217**(24), 4365–4371 (2014)

Part III
Feasible Actions of Tendon-Driven Limbs

Neuromechanics requires understanding how the command signals from the nervous system are transformed into motor actions. This Part focuses on identifying and characterizing the feasible actions a tendon-driven limb can produce. After all, the nervous system can only learn to exploit the mechanical capabilities of the physical system it controls.

Chapter 7
Feasible Neural Commands and Feasible Mechanical Outputs

Abstract Having understood the basic concepts related to the structure and function of tendon-driven limbs, we can now revisit and appreciate the question: 'How does the nervous system control a tendon-driven limb?' Chap. 5 shows why it is reasonable to consider optimization as a means to resolve the muscle redundancy that exists for the control of joint torques. But Chap. 6 paints an alternative picture when we see that orchestrating tendon excursions among muscles is severely overdetermined—which is the opposite of redundancy. In this chapter I begin to explore in detail the working hypothesis that having numerous muscles does not make them as redundant as proposed by the classical notion of muscle redundancy. I do this by introducing you to the concepts of how feasible muscle activations produce feasible mechanical outputs. This perspective grows out of the fusion of linear algebra, geometry, mechanics, and anatomy. It shows how the anatomical structure of the limb together with the constraints that define a mechanical task naturally specify a set of feasible neural commands. The fact that this set of feasible neural commands has a well-defined structure compels and allows us to revise and extend the classical notion of muscle redundancy, and propose a more general approach to neuromuscular control that emphasizes compatibility with evolutionary biology and clinical reality.

7.1 Mapping from Neural Commands to Mechanical Outputs

Let us revisit the underdetermined problem of muscle redundancy by presenting a graphical interpretation as in [1, 2], which is an extension and generalization of prior graphical interpretations [3–6].[1] The problem is phrased in its general form in Eq. 5.7, presented here as Eq. 7.1. Figure 7.1 shows an example for a planar system with 2 DOFs and 3 muscles

$$\mathbf{w} = J^{-T} R F_0 \mathbf{a} = H \mathbf{a} \qquad (7.1)$$

---

[1] I was first introduced to these concepts by Art Kuo, a fellow doctoral student who showed me how neural commands produce feasible mechanical actions in the context of limb accelerations [6].

© Springer-Verlag London 2016   91

F.J. Valero-Cuevas, *Fundamentals of Neuromechanics*,
Biosystems & Biorobotics 8, DOI 10.1007/978-1-4471-6747-1_7

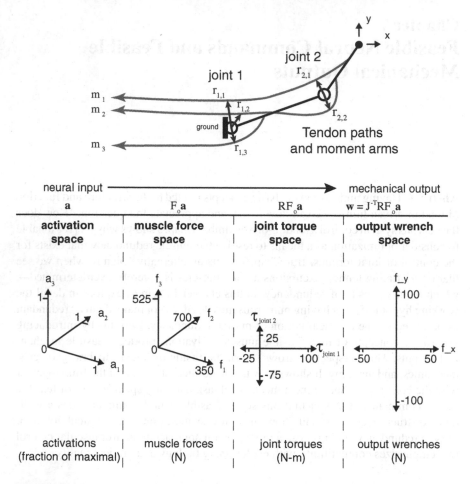

Fig. 7.1 Relationship between matrix multiplication and mapping between successive vector spaces. This forward transformation from neural input (activation vector **a**) to mechanical output (wrench vector **w**) is also a mapping of vectors from activation space to endpoint wrench space with its associated changes in units and dimensionality. Adapted with permission from [2]

Note that matrix multiplication in Eq. 7.1 proceeds from right to left, and Fig. 7.1 presents the operations from left to right, as is customary to show causal progressions. This corresponds to:

(a) begin by using the graphical interpretation of the vector space $\mathbf{a} \in \mathbb{R}^3$, which is a 3D Cartesian space,[2] where each independently controlled muscle is a dimension that corresponds to each of the axes. This *activation space* in $\mathbb{R}^3$ is the neural

---

[2]Using 3 muscles in this example is useful to facilitate visualization of the 3D activation space, but the concepts here extend to any number of muscles in higher dimensions [11, 12]. See also Appendix A.

input to the limb for this formulation. The values that each element of **a** can assume are between 0 and 1. A point (i.e., a vector **a**) in activation space is then *mapped* into points in

(b) *muscle force space*, also in $\mathbb{R}^3$, by the diagonal matrix F_0 from Eq. 5.3. The matrix R then maps points in muscle space into points in

(c) *joint torque space* that is in $\mathbb{R}^2$. This reduction in dimensionality happens because there are typically more muscles than kinematics DOFs. Last, points in joint torque space are mapped by the matrix J^{-T} into points in

(d) *endpoint wrench space*. In this case, that wrench endpoint space is also 2D as it includes only the forces f_x and f_y. This is the mechanical output of the limb for this formulation.

Generally:

- The dimensionality of the activation space is that of the vector $\mathbf{a} \in \mathbb{R}^N$, where N is the number of independently controlled muscles.
- The dimensionality of the muscle force space is also $\mathbf{f}_m \in \mathbb{R}^N$ as in Eq. 4.12.
- The dimensionality of the joint torque space is that of the vector $\boldsymbol{\tau} \in \mathbb{R}^M$ as in Eq. 4.24, where M is the number of kinematic DOFs. This is as in Eq. 5.8 specified by the geometric model that was selected.
- If the number of kinematic DOFs $M > 6$, the dimensionality of the endpoint wrench space is still 6 (because a rigid body only has 6 kinematic DOFs), but then you have kinematic redundancy where multiple postures of the limb can produce the same position *and* orientation of the endpoint, much as in snake robots [7].[3]

When thinking about the nature and structure of each matrix in terms of mapping across *vector spaces* as shown in Fig. 7.1, we should note that:

- The matrix F_0 is diagonal. This means that it scales each element of the activation vector **a** to transform it into a vector of muscle forces $\mathbf{f}_m$. The mapping is from $\mathbb{R}^N$ to $\mathbb{R}^N$.
- The matrix $R \in \mathbb{R}^{M \times N}$ maps vectors $\mathbf{f}_m$ from the higher dimensional muscle force space $\mathbb{R}^N$ into vectors $\boldsymbol{\tau}$ in the lower dimensional joint torque space $\mathbb{R}^M$.
- The matrix J^{-T} maps vectors $\boldsymbol{\tau}$ in joint torque space into vectors **w** in endpoint wrench space. If the Jacobian J is full rank, it is usually $\in \mathbb{R}^{M \times M}$ and it is invertible, as discussed in relation to Eq. 5.8. This happens when the internal and external kinematic DOFs have the same dimensionality (see Sects. 2.6 and 3.5) and J^{-T} can be found exactly as it would represent conservation of energy (see

---

[3] A note about kinematic redundancy in the workspace of the endpoint of the limb. You can have some kinematic redundancy for, say, the endpoint location already with fewer than 6 kinematic DOFs. An example is when more than 3 axes of rotation are all aligned and lie on the same plane. Even when this is not the case, when you can have a 'double-jointed' finger or elbow because two mirror-image postures achieve the same endpoint location or orientation—but not both. Thus kinematic redundancy also depends on the geometric model you choose. To have full kinematic redundancy where you can obtain the same position *and* orientation anywhere in space, you need more than 6 well-chosen kinematic DOFs, as with a snake robot. Note that kinematic redundancy also implies kinetic redundancy in the wrench space of the limb.

its derivation in Chap. 3). Section 3.7 discusses cases where the Jacobian is not invertible.

Let us give these concepts a physiological interpretation. What the mappings in Fig. 7.1 say is that every neuromuscular system has a particular neural wherewithal (i.e., set of feasible neural commands) that naturally leads to a particular mechanical wherewithal (i.e., set of feasible mechanical outputs). Typically the set of feasible neural commands is 'larger' (i.e., of a higher dimension) than the space of feasible mechanical actions. More specifically, this leads to posing the mapping from neural input to net joint torques as the underdetermined problem that is at the core of the classical notion of muscle redundancy. It is then reasonable to think that the nervous system has multiple choices to produce a given joint torque output.

Therefore, the purpose of a computational geometry approach to muscle redundancy is to understand the structure of the feasible neural commands for real world tasks. The geometric interpretation of mapping from activation space to endpoint wrench space allows us to see that the structure of the set of feasible neural commands itself imposes limits on—and correlations among—the feasible commands to each muscle. It also opens new avenues of thinking about motor learning and motor performance that go beyond the current emphases on optimization and cost functions.

This chapter is dedicated to showing how the structure of feasible neural commands comes about. In essence, the anatomy of the limb and the constraints that define the functional requirements of real-world ecological behavior (which can be mechanical, metabolic, physiological, etc.) naturally limit and give structure to the set of feasible neural commands. There are many strong connections between this computational geometry approach and linear programming presented in Sect. 5.3. In fact, in Chap. 8, I will describe how these two approaches can be considered to be mathematical counterparts (i.e., duals of each other). But most importantly, this chapter highlights that all the nervous system can do is use (i.e., learn, explore, and exploit) the set of feasible neural commands to produce the set of feasible mechanical outputs of the physical system it controls. This perspective allows us to see muscle redundancy in a different light.

7.2 Geometric Interpretation of Feasibility

The connection among linear algebra, optimization, and geometry is a deep one. One can argue that they are one and the same. This is why it is important to note the very specific relationships between the graphical interpretation of these mappings and the structure of the equations that produce them. Parts I and II of this book emphasized the creation of a mathematical description of the neuromuscular actuation and control of tendon-driven limbs. Here we use those mathematical descriptions to understand the set of feasible neural commands and mechanical outputs.

In particular, matrix equations form the basis of the mathematical descriptions we use for the underdetermined case of static force production, and the overdetermined case of tendon excursions. Then let us begin by introducing the concept of the *column space* of a matrix.

In linear algebra [8], the range of a matrix A is its column space $\mathscr{C}(A)$, which is the set of all possible linear combinations of its column vectors. That is, the product of a matrix A with a column vector $\mathbf{x}$ can be written as the linear combination of the column vectors of the matrix. The coefficients of that linear combination are the elements of the vector $\mathbf{x}$. For matrix $A \in \mathbb{R}^{M \times N}$ with N column vectors $\mathbf{v}_i \in \mathbb{R}^M$, $i = 1, \ldots, N$, multiplied by a vector $\mathbf{x} \in \mathbb{R}^N$,

$$\mathbf{y} = A\,\mathbf{x} = \begin{bmatrix} \mathbf{v}_1\ \mathbf{v}_2 \ldots \mathbf{v}_N \end{bmatrix} \begin{pmatrix} x_1 \\ x_2 \\ \vdots \\ x_N \end{pmatrix} = x_1\,\mathbf{v}_1 + x_2\,\mathbf{v}_2 + \cdots + x_N\,\mathbf{v}_N \qquad (7.2)$$

where $x_1, x_2, \ldots, x_N$ are scalars. The set of all possible linear combinations of $\mathbf{v}_1$, $\mathbf{v}_2, \ldots, \mathbf{v}_N$ is called the column space of A. And the column space of A is also the *span* of the vectors $\mathbf{v}_1, \mathbf{v}_2, \ldots, \mathbf{v}_N$. The word 'span' has the very intuitive meaning of the regions that can be reached by combining the different columns of a matrix.

Said differently, given the matrix operation $\mathbf{y} = A\mathbf{x}$ that maps from $\mathbb{R}^N$ to $\mathbb{R}^M$, any 'output' vector $\mathbf{y}$ can only exist in the column space of A

$$\mathbf{y} \in \mathscr{C}(A) \qquad (7.3)$$

which means that all *feasible outputs* are explicitly defined by the column space of A.

What is the dimensionality, size and shape of a column space? Well, that depends on both A and $\mathbf{x}$, of course. The dimensionality of the column space is $\mathbb{R}^M$ if matrix A is of rank M. Intuitively, if the column vectors of A are all replicas of each other, then the column space is simply the line represented by them—and its dimension is 1. But if all column vectors are linearly independent, then in principle their span is of dimensionality M, i.e., $\mathscr{C}(A) \in \mathbb{R}^M$, if $M \leq N$.

The size of $\mathscr{C}(A)$ depends on whether or not the scalars that form $\mathbf{x}$ are bounded. If their values are not bounded, then the span of A stretches out to infinity in the space $\mathbb{R}^M$.

Without going into too much detail, if the elements of $\mathbf{x}$ are bounded

$$\begin{aligned} x_1\,\mathbf{v}_1 + x_2\,\mathbf{v}_2 + \cdots + x_N\,\mathbf{v}_N \\ a_i \leq x_i \leq b_i \\ -\infty < a_i \leq b_i < \infty, \; i = 1, \ldots, N \end{aligned} \qquad (7.4)$$

then the column space $\mathscr{C}(A)$ will also be bounded. Different choices of bounds a and b produce different collections of points.

The connection to geometry is clear here as you can see that these collections of points are subspaces in the shape of a *polygon, polyhedron,* or *polytope* for $M = 2, 3$ and higher dimensions, respectively.

In elementary geometry, a polytope is a geometric object with flat sides, and may exist in any general number of dimensions N as an N-dimensional polytope or N-polytope. For example a 2D polygon is a 2-polytope and a 3D polyhedron is a 3-polytope. For simplicity I will use the general term polytope instead of N-polytope, and polygon and polyhedron as needed [9].

Let us be more specific to the topic of control of tendon-driven limbs. Here we have the strong constraint that muscles can only be commanded to pull, but not push, on their tendons as indicated in Eqs. 5.1 and 5.2. Thus the admissible linear combinations are

$$x_1 \, \mathbf{v}_1 + x_2 \, \mathbf{v}_2 + \cdots + x_N \, \mathbf{v}_N$$
$$0 \le x_i \le 1, \; i = 1, \ldots, N \tag{7.5}$$

which applies to all the matrix multiplications in Fig. 7.1.

Figure 7.2 shows the convex polygon that is the column space created by such bounded input. When you perform all possible positive combinations of the sums of vectors, where the scalars x_i are bounded between 0 and 1, it is called a *Minkowski sum* [10]. Note that a Minkowski sum produces many points in the shape of a polygon.

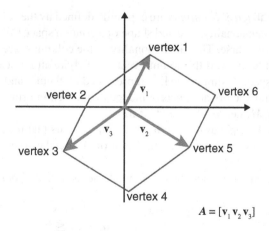

$$A = [\mathbf{v}_1 \, \mathbf{v}_2 \, \mathbf{v}_3]$$

Fig. 7.2 A zonotope is a subspace created by the *positive* linear combination of vectors. A Minkowski sum, see Eq. 7.5, is a special case where the scalar weights of the linear combination lie between 0 and 1. For example, the Minkowski sum of the column vectors of matrix $A = [\mathbf{v}_1 \, \mathbf{v}_2 \, \mathbf{v}_3]$, where $A \in \mathbb{R}^{2 \times 3}$, can be represented graphically as the positive sum of line segments connecting the origin to the endpoint of each vector $\mathbf{v}_i$. The column vectors $\mathbf{v}_i$ are called *generators* because they are the building blocks that produce the zonotope

But if you shrink-wrap all points to find only those that are the vertices of the polygon that contains all points, you find the convex hull of the Minkowski sum. To see how this is done consider the Minkowski sum seen in Fig. 7.2, where the vectors $\mathbf{v}_i$ are called *generators* because they are the building blocks of the subspace.

Here we see that the Minkowski sum produces a convex polygon with 6 vertices:

$$
\begin{aligned}
\text{vertex } 1 &= 1\,\mathbf{v}_1 + 0\,\mathbf{v}_2 + 0\,\mathbf{v}_3 \\
\text{vertex } 2 &= 1\,\mathbf{v}_1 + 0\,\mathbf{v}_2 + 1\,\mathbf{v}_3 \\
\text{vertex } 3 &= 0\,\mathbf{v}_1 + 0\,\mathbf{v}_2 + 1\,\mathbf{v}_3 \\
\text{vertex } 4 &= 0\,\mathbf{v}_1 + 1\,\mathbf{v}_2 + 1\,\mathbf{v}_3 \\
\text{vertex } 5 &= 0\,\mathbf{v}_1 + 1\,\mathbf{v}_2 + 0\,\mathbf{v}_3 \\
\text{vertex } 6 &= 1\,\mathbf{v}_1 + 1\,\mathbf{v}_2 + 0\,\mathbf{v}_3
\end{aligned}
\tag{7.6}
$$

But notice that, for example, the sum $1\,\mathbf{v}_1 + 1\,\mathbf{v}_2 + 1\,\mathbf{v}_3$ is not a vertex. Why? See Sect. 7.6 to explore this further.

Such polygons are called a *zonotopes* in 2D, and *zonohedra* (singular *zonohedron*) in 3D or higher dimensions [11, 12]). The fact that they are convex by construction (see Sect. 7.6) has many important consequences we will see in subsequent chapters. And any convex polygon can be described and defined equivalently by either its vertices, or the linear equations that define its sides [13]. At this point the ideas of convexity and convex sets become important. Please see Sect. 7.6 where these are developed further.

7.3 Introduction to Feasible Sets

We now apply Eq. 7.3 to tendon-driven limbs to find the set of all feasible mechanical outputs of the system in terms of joint torques and endpoint wrenches. Once again, the nervous system can only produce positive linear combinations of the column vectors of matrix F_0, matrix $[R\ F_0]$, and matrix $H = [J^{-T}\ R\ F_0]$ as shown in Fig. 7.1. Therefore, as shown in Fig. 7.3:

- The set of all possible muscle force vectors is the column space of the matrix F_0. This is called the *feasible muscle force set*. This zonotope (or zonohedron) is produced by the Minkowski sum of all possible muscle force capabilities of the limb. That is, the column vectors of F_0 are the generators the nervous system has at its disposal to produce muscle forces.
- The set of all possible joint torque vectors is the column space of the matrix $[R\ F_0]$. This is called the *feasible torque set*. This zonotope (or zonohedron) is produced by the Minkowski sum of all possible torque capabilities of the

limb. That is, the column vectors of $[R\ F_0]$ are the generators the nervous system has at its disposal to produce joint torques.

- The set of all possible output wrench vectors is the column space of the matrix $H = [J^{-T}\ R\ F_0]$. This is called the *feasible wrench set*. This zonotope (or zonohedron) is produced by the Minkowski sum of all possible wrench capabilities of the limb. That is, the column vectors of $H = [J^{-T}\ R\ F_0]$ are the generators the nervous system has at its disposal to produce output wrenches.

- Note also that the set of all feasible activations is naturally a convex polyhedron, a cube with sides of unit length that exists in the positive octant[4] of the N-dimensional activation space (or a *positive unit N-cube*, which is also a zonohedron (why?). This is the *feasible activation set* [1]. Thus all other feasible sets are simply the mapping of the feasible activation space by the matrices in Fig. 7.1. More on this later in this chapter.

To find these feasible spaces, we use the column vectors of these matrices. For example, for a limb with M kinematic DOFs (joints) and N independently controlled muscles, the transformation

$$[R\ F_0]\ \mathbf{a} \tag{7.7}$$

takes into account the first two steps in Fig. 7.1. The column space of this transformation is readily obtained if we multiply R by the diagonal matrix F_0

$$[R\ F_0] \in \mathbb{R}^{M \times N} \tag{7.8}$$

$$\boldsymbol{\tau} = R\ F_0\ \mathbf{a} = \begin{bmatrix} \mathbf{r}_1\ \mathbf{r}_2\ \dots\ \mathbf{r}_N \end{bmatrix} \begin{bmatrix} F_1 & & & \\ & F_2 & & \\ & & \ddots & \\ & & & F_N \end{bmatrix} \mathbf{a} \tag{7.9}$$

$$\boldsymbol{\tau} = \begin{bmatrix} F_1\ \mathbf{r}_1\ F_2\ \mathbf{r}_2\ \dots\ F_N\ \mathbf{r}_N \end{bmatrix} \mathbf{a} \tag{7.10}$$

Note that

$$F_i\ \mathbf{r}_i = \begin{pmatrix} \tau_1 \\ \tau_2 \\ \vdots \\ \tau_M \end{pmatrix} \tag{7.11}$$

is simply the vector that contains the maximal torques that muscle i produces at every joint. And thus

---

[4] An octant is the generalization of a quadrant to Cartesian spaces of dimensionality ≥ 3.

$$\boldsymbol{\tau}_i = F_i \mathbf{r}_i \tag{7.12}$$

which means that

$$\boldsymbol{\tau} = \begin{bmatrix} \boldsymbol{\tau}_1 \ \boldsymbol{\tau}_2 \ \dots \ \boldsymbol{\tau}_N \end{bmatrix} \begin{pmatrix} a_1 \\ a_2 \\ \vdots \\ a_N \end{pmatrix} \tag{7.13}$$

Equation 7.13 is particularly revealing. Comparing it to Eq. 7.2 we see that the columns of matrix $[R \ F_0]$, each $\boldsymbol{\tau}_i$, are the 'generators' of the Minkowski sum in Fig. 7.2. Therefore, the zonotope they create is the *feasible torque set* because it describes and contains all possible net joint torques the nervous system can produce. Said differently, the neural command $\mathbf{a}$ defines how the nervous system combines the actions of the N muscles to produce the net joint torques $\boldsymbol{\tau} \in \mathbb{R}^M$. And the feasible torque set, therefore, shows all possible net joint torques the nervous system can possibly produce in that joint torque space.

Similarly, the mapping

$$[J^{-T} R \ F_0]\mathbf{a} \tag{7.14}$$

takes into account all three steps in the mapping from activation space to wrench space as shown in Fig. 7.1. The column space of this transformation is readily obtained if we define the column vectors of the matrix H as $\mathbf{w}_i \in \mathbb{R}^M$ (assuming that the output wrench space is M-dimensional, but it could be otherwise as per the geometric model of the limb)

$$H = [J^{-T} R \ F_0] \in \mathbb{R}^{M \times N} \tag{7.15}$$

$$\mathbf{w} = J^{-T} R \ F_0 \mathbf{a} = \begin{bmatrix} \mathbf{w}_1 \ \mathbf{w}_2 \ \dots \ \mathbf{w}_N \end{bmatrix} \begin{pmatrix} a_1 \\ a_2 \\ \vdots \\ a_N \end{pmatrix} \tag{7.16}$$

Likewise, the column vectors $\mathbf{w}_i$ of matrix $[J^{-T} \ R \ F_0]$ are the generators of the Minkowski sum in Fig. 7.2. Therefore, the zonotope they create is the *feasible wrench set* because it describes and contains all possible wrenches the nervous system can produce at the endpoint of the limb in that particular posture. Each generator $\mathbf{w}_i$ is the wrench vector muscle i produces, and the neural command $\mathbf{a}$ defines how the nervous system combines the actions of the N muscles to produce the output wrench $\mathbf{w}$.

To summarize, Fig. 7.3 starts with the cube that is the most general feasible activation set, which first creates the parallelepiped of the feasible muscle force set, and then the zonotopes of the feasible joint torque, and feasible output wrench sets. Note that the column vectors $\boldsymbol{\tau}_i$ and $\mathbf{w}_i$ are in the units of joint torques and end-

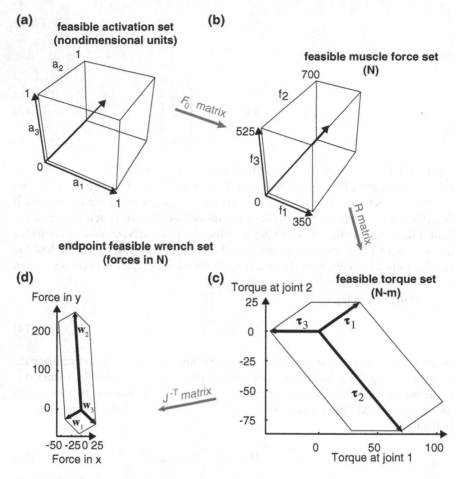

Fig. 7.3 Sequence of linear transformations of the feasible activation set in a forward model of static force production by a tendon-driven limb. Using the concepts in Figs. 7.1 and 7.2, we use as neural input **a** the feasible activation set (a positive unit cube) to produce **b** the *feasible muscle*, **c** *feasible joint torque*, and **d** *endpoint feasible wrench* sets. Adapted with permission from [2]

point wrenches, respectively. Their positive linear combinations specify the set of all possible outputs in those spaces. If we consider them as the $\mathbf{v}_i$ vectors of Fig. 7.2, we can build the zonotopes that correspond to the *feasible joint torque set* and the *feasible output wrench set* as shown in Fig. 7.3. In the specific simple case shown in Fig. 7.4, the endpoint wrench is a planar force vector—consisting of f_x and f_y components only. This zonotope is called the *feasible force set*, and not the feasible wrench set because the endpoint output contains no torques (see Sect. 2.6).

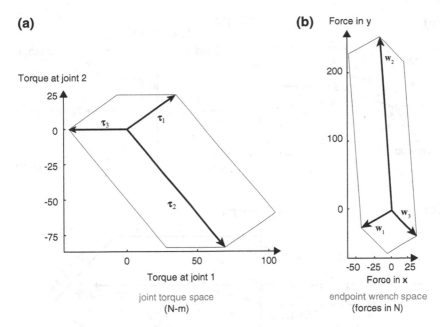

Fig. 7.4 Zonotopes representing the *feasible torque set* (*left*) and the *feasible wrench set* (*right*). They are obtained from the column vectors defined in Eqs. 7.13 and 7.16. Given that endpoint wrench in this example is planar force, the feasible wrench set is actually a *feasible force set* as it only contains the output force components f_x and f_y (i.e., no torque components, see Sect. 2.6). Note these unconstrained feasible sets are both zonotopes because they are created purely by the Minkowski sum of the generator vectors as in Fig. 7.2. Adapted with permission from [2]

7.4   Calculating Feasible Sets for Tasks with No Functional Constraints

Note that we are studying the general case of the unconstrained feasible torque, wrench, and force sets, as shown in Fig. 7.3. I call these 'unconstrained' feasible sets because they describe what the system can do when you simply combine all possible actions without regard to functional constraints or goals, such as canceling out some elements of the output wrench **w** while keeping others. The only *a priori* constraints are that the activation, **a**, to all muscles be limited to be ≥ 0 and ≤ 1. Thus the feasible sets are all zonotopes because they are created by the Minkowski sum of the generator vectors as in Fig. 7.2. For now let us stay with this most general case to discuss the calculation and properties of such convex sets for neuromechanical function.

Calculating these unconstrained feasible sets is quite straightforward and can be done in two equivalent ways:

- The first is to perform the Minkowski sum of the relevant column vectors, as we saw in Figs. 7.2, 7.3, and 7.4. This produces many points, only a few of which

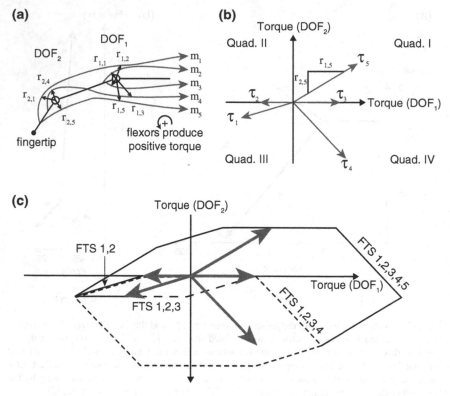

Fig. 7.5 Example of how to create the feasible torque set (FTS) for a planar 5 muscle system. Begin by finding all possible linear combinations of column vectors τ_1 and τ_2. This is simply done by using vector addition with the parallelogram method, and is labeled FTS 1, 2. When adding τ_3, simply add that vector to the vertices of FTS 1, 2 to obtain FTS 1, 2, 3. Repeat the process to obtain the full feasible torque set FTS 1, 2, 3, 4, 5. Adapted with permission from [14]

actually become vertices of the convex hull. See Fig. 7.5 for an example that uses a planar 5 muscle system.

- Alternatively, we can think of these unconstrained feasible sets as the mapping of convex sets from one space to another as shown in Fig. 7.3.

The second approach hinges on the fact that the neural input to the system—the set of feasible activations—is bounded to the positive octant in activation space. That is, if $0 \leq \mathbf{a} \leq 1$, then its elements are also bounded as $0 \leq a_i \leq 1$ for $i = 1, \ldots, N$. This *feasible activation space* is a unit positive N-cube in $\mathbb{R}^N$, as shown in Fig. 7.3. An N-cube is the generalization of a cube to >3 dimensions. It is called an N-cube because it is N-dimensional. And it is a 'unit positive' N-cube because all of its sides have length 1, with values along each dimension ranging from 0 to positive 1. Such a non-negative convex set then lies in the positive octant of $\mathbb{R}^N$. An octant is the generalization of a Cartesian quadrant to 3D where there are 8 such regions. The 8 vertices of the cube representing the unit positive octant are, in no particular order,

$$\begin{pmatrix}0\\0\\0\end{pmatrix}, \begin{pmatrix}1\\0\\0\end{pmatrix}, \begin{pmatrix}0\\1\\0\end{pmatrix}, \begin{pmatrix}0\\0\\1\end{pmatrix}, \begin{pmatrix}1\\1\\0\end{pmatrix}, \begin{pmatrix}1\\0\\1\end{pmatrix}, \begin{pmatrix}0\\1\\1\end{pmatrix}, \begin{pmatrix}1\\1\\1\end{pmatrix} \qquad (7.17)$$

Now think of the mapping that F_0 performs. This operation takes every point in the feasible activation space (a vector $\mathbf{a}$ in the unit positive N-cube) and transforms it into a point in muscle force space (a vector $\mathbf{f}_m$). Note that the feasible activation space is a convex set (an N-cube that is by definition convex), and that the matrix F_0 performs a linear mapping. Because convex sets remain convex under linear mapping [13, 15], then the *feasible muscle force set* will be a convex set as well. It is in fact a parallelepiped, which is a cube whose sides are stretched differently by the diagonal elements in the matrix F_0.

Similarly, mapping the feasible muscle force set by the matrix R will be a convex set in joint torque space: the feasible torque space. And mapping the feasible torque space by the J^{-T} will also produce a convex set in output wrench space: the feasible wrench set. Once again, this sequence is shown in Fig. 7.3.

Figure 7.3 merits some explanation as it holds important lessons. The initial mapping from activation space to muscle force space happens within a similar dimensionality, namely $\mathbb{R}^3$. Thus, a diagonal matrix distorts a cube into a parallelepiped.

Now consider the mapping from $\mathbb{R}^3$ to $\mathbb{R}^2$. Such mapping is needed in Fig. 7.3 to find the feasible torque set. This is an operation analogous to casting a shadow as shown in Fig. 7.6, and is further described in Sect. 7.6.

- The extreme points of the projected convex set are a function of the extreme points of the original convex set. That is, all edges of the shadow are produced by edges, facets, and vertices of the cube.

  – Therefore, every extreme point of the shadow is produced by mapping a unique point on the surface of the original object. See the vertices of the shadow numbered 1, 2, 3, 6, 7, and 8 in Fig. 7.6. This is in case of no degeneracies— which we assume. Can you think of a degenerate case, though?
  – This means that producing a maximal muscle force, joint torque, or output wrench vector (i.e., a point on the boundary of a feasible set) can only be done by a unique muscle activation pattern $\mathbf{a}^*$ [1]. Figure 7.7 shows two such examples.

- The internal points of the projected convex set can be a function of both internal or extreme points of the original convex set. That is, the points internal to the shadow are produced by internal points of the cube, and edges, facets, and vertices of the cube.

  – Therefore, every internal point of the shadow can be produced by mapping an infinite number of points of the original object (i.e., those that lie on the solid line that goes through the original object and projects onto an internal point). See point 4 in Fig. 7.6. The points on the solid line (i.e., the solution space) are the *null space* of the output because, even though they are different, they produce no measurable change in their shadow (i.e., a same point).

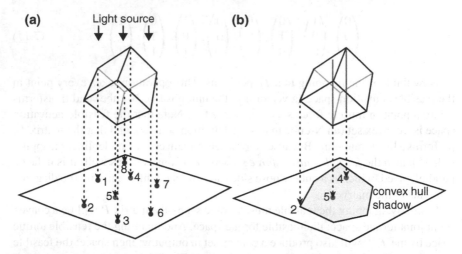

Fig. 7.6 Mapping a convex object from a higher (i.e., a *cube*) to a lower (i.e., a *polygonal shadow*) dimension. A convex object remains convex under linear mapping, as when casting a shadow from a 3D object onto 2D. In this example a convex polyhedron in 3D, a cube, maps into a convex polygon in 2D, its shadow. **a** One way to find the lower-dimensional projection of a convex polyhedron is to project each of its vertices onto the lower dimension, and find those that become extreme points in that lower dimension (vertices 1, 2, 3, 6, 7, and 8). **b** These extreme points define a convex hull that is a *polygonal shadow*. But some extreme points of the higher-dimensional object become internal points of the lower-dimensional object (points 4 and 5). Yet every point along the boundary of the lower-dimensional object is (often) produced by a unique point on the edges of the high-dimensional object (why?). In contrast, all points internal to the lower-dimensional object are the projection of extreme points and an infinite number of internal points of the higher-dimensional object. See Sects. 7.6 and 8.3 for details

- This means that producing a sub-maximal muscle force, joint torque, or wrench (i.e., an output in the interior of a feasible set) can be done by an infinite number of muscle activation patterns [1]. This is one definition of muscle redundancy for net joint torque production.

How do we find the coordination pattern that produces a particular maximal output vector? The muscle activation pattern that produces each point on the boundary of the feasible force set is simply the extreme point in the N-cube that produces that point. This is shown graphically in Fig. 7.6. If the intersection of the ray and the boundary of the feasible force is a vertex, it is produced by a vertex of the N-cube (why?). If the intersection of the ray and the boundary of the feasible force set is on one of the sides, then the corresponding coordination pattern is the weighted average of the coordination patterns that produce the vertices at either end of that side [1].

Figure 7.8 shows the maximal forces vectors in the A, B, and C directions, $\mathbf{f}_A$, $\mathbf{f}_B$, and $\mathbf{f}_C$, which are produced by the optimal coordination patterns $\mathbf{a}_A^*$, $\mathbf{a}_B^*$, and $\mathbf{a}_C^*$, respectively. The coordination patterns $\mathbf{a}_A^*$ and $\mathbf{a}_C^*$ are obtained from the corresponding vertices of the N-cube. But $\mathbf{a}_B^*$ can be obtained as

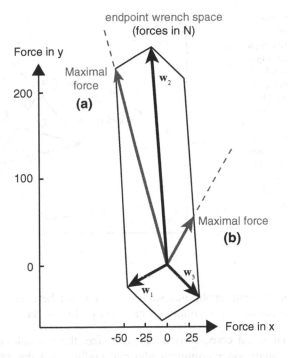

Fig. 7.7 Finding the maximal possible force in a given direction. In this case, the feasible output wrench set only consists of forces in the x and y directions. Thus it is a feasible output force set, not a feasible output wrench set. A ray from the origin in a given direction of interest will intercept the boundary of the feasible force set. The length of that vector is the magnitude of the maximal forces in that direction. The unique activation pattern that produces a maximal output is the point on the surface of the feasible activation set that produces that point on the boundary of the feasible force set. In the simplest case, **a** the force direction passes through a vertex of the feasible output set, and the coordination pattern that produces it is that vertex in the feasible activation set that corresponds to it. Alternatively, **b** the maximal force passes though the boundary and the coordination pattern that produces it is the weighted sum of the vertices in the feasible activation set that produce the adjacent vertices in the feasible force set, see Eq. 7.18 and Fig. 7.8

$$\mathbf{a}_B^* = \frac{d_2\, \mathbf{a}_A^* + d_1\, \mathbf{a}_C^*}{d_1 + d_2} \qquad (7.18)$$

7.5 Size and Shape of Feasible Sets

The first thing to notice when you build the feasible joint torque set in Fig. 7.5 is that each and every muscle affects the size and shape of the zonotope. Given that the transformation by the J^{-T} is linear, it stands to reason that the larger the feasible joint torque set is, the larger the feasible output wrench set will be. Having large joint torque and feasible force sets is, of course, desirable as it grants the system

Fig. 7.8 The activation patterns that produce maximal output are unique. In this example showing a feasible force set, each vertex and point along the boundary is produced by a unique muscle activation pattern. These are either the vertices of the feasible activation set that produce the vertices of the feasible force set (e.g., $\mathbf{a}_A^*$ and $\mathbf{a}_C^*$), or their weighted sum (e.g., $\mathbf{a}_B^*$) as per Eq. 7.18

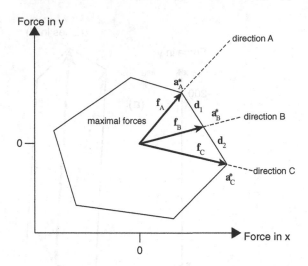

greater mechanical output capabilities, such as strength, but there are some interesting physiological consequences to zonotopes created by Minkowski sums:

- Given the anatomical constraints on muscle size, the physiological regulation of force magnitude via recruitment and rate coding, and the nature of motor noise, having more smaller muscles may be more desirable than having few larger muscles.
- Compare Figs. 7.5 and 7.9. The latter was produced using the script `zonotope` (available as indicated in Sect. 7.7) with 3, 4, 5, 6, 7, or 8 generators (i.e., independent muscles). Note that simply adding more muscles is not necessarily sufficient to improve mechanical output in general. More muscles will inevitably grow the size of the zonotope, but its shape may or may not benefit from that. In the case of Fig. 7.5, the addition of more muscles grows the feasible joint torque set in ways where each quadrant benefits. In the case of the random matrix H, Fig. 7.9 shows that the addition of muscles does not increase the reach of the zonotope into

Fig. 7.9 Zonotope created to simulate the addition of 3 to 8 generators (i.e., independently controlled muscles). This was done by defining a random matrix $H \in \mathbb{R}^{2 \times 8}$ and then using its first 3 to 8 column vectors, which is the same as adding the 3rd to 8th generator, one generator at a time

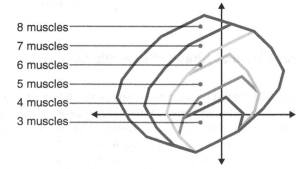

the 4th (i.e., lower right) quadrant. Thus adding musculature whose tendon paths lead to growth of the zonotope into desired regions or quadrants is an important consideration.

- But there are counterarguments to suggesting that limbs should evolve to span all quadrants (or octants in higher dimensions) of their output spaces equally. We have shown that the human index finger and thumb have asymmetric force capabilities in directions associated with grasp forces [1, 16–18]. Another recent study showed that the tendon routing and distribution of size across the muscles of the index finger and thumb produce asymmetric output capabilities that are particularly well adapted to producing forces for grasp [19]. This holds important lessons to the engineering design of robotic limbs driven by tendons.
- Tendon routing—be it well-adapted in biological systems, or judiciously chosen by engineers in robotic systems—has profound consequences to function, yet tendon routing has remained relatively unexplored in robotics as described in [20]. In contrast, hand surgeons have debated for decades the proper routing of tendons for reconstructive surgeries. Many of these surgeries, called *tendon transfers* [5, 21–23], began to be developed in their current form in the early 20th century in India by Richard Brand to compensate for the functional deficits induced by paralysis of hand muscles by Hansen's disease (leprosy). When survival rates for spinal cord injury began to improve in the 1950s and 60s, surgeons like Brand, Littler, and Moberg extended the use of these techniques into what are now the standard procedures as described, for example, by Hentz and Leclercq [22]. In fact, this is where diagrams of each muscle's action in *joint torque space* first appeared, and became popular when designing tendon transfer surgeries.

7.6   Anatomy of a Convex Polygon, Polyhedron, and Polytope

In Chap. 8 we will continue to explore important properties of feasible sets as they relate to neural control. But it is already relevant to point out some important features of convex sets. In this book I take convex sets to mean convex and bounded polygons, polyhedra and polytopes as defined in Sect. 7.2. I will at times use all terms interchangeably because the properties of bounded convex sets usually cut across dimensions when we are looking at the specific case of systems of linear equations and linear inequality constraints. Just above I mentioned that convex sets under linear mapping remain convex, thus the mappings of the feasible activation set will produce convex feasible joint torque and feasible force sets. What does this mean?

Consider Fig. 7.6 where it is clear that the shadow a cube casts onto a 2D plane is a convex polygon. A linear operation from a higher to a lower dimension is exactly like casting a shadow where light rays are lines that project the entire boundary of the high-dimensional object onto a lower dimension. It is clear to see how the shadow of a convex object—a cube—will itself be convex. It is also clear that the properties of the matrix that performs the mapping define the source of the light rays, how they shine around the object, and therefore the dimensionality and shape of the shadow they will cast. Section 7.7 has computer code to implement the mapping of an N-cube of dimensions 3 or higher onto the 2D or 3D Cartesian spaces.

Recall that some of the vertices of the cube get projected to the inside of the shadow. Therefore, when mapping a convex object from a higher to a lower dimension, the whole column of points of a light ray that passes through the interior of the cube produces a single point in the interior of the shadow (see the solid lines that project onto vertices 4 and 5 in Fig. 7.6). In contrast, those points that make the boundary shadow are produced by the light rays that glance a single point in the boundary of the cube. There are rare exceptions, of course, such as when the light is cast in such a way that an edge of the 3D object is exactly parallel to the light rays. This results in a vertex on the boundary of the shadow that is in fact the projection of the entire set of points that form that edge of the 3D object.

As to the anatomy of a convex set, notice that in Fig. 7.6 you can easily and quickly find the boundary of the shadow by finding how the vertices of the cube map onto the lower-dimensional space. By this I mean that if you were to shrink-wrap a circle that originally contains all of the projected vertices of the cube, you would create a polygonal shadow whose straight lines would connect the most external points. This polygon is called the *convex hull* of the set of points. Similarly, the convex hull operation is that of shrink-wrapping a group of points to find the convex set that includes them all. Each of the extreme points becomes a *vertex* of the polygon.

Therefore, there are two equivalent ways to represent and define a convex polygon, polyhedron or polytope:

- *Vertex representation*: In the case of *bounded* polygons, polyhedra and polytopes, I simply use an ordered list of the vertices of the polygon, polyhedron, or polytope, and connect them with straight lines, planes, or hyperplanes to form the facets of the convex and bounded set, respectively.[5] For example, a square can be described by the vector coordinates of its 4 vertices as

$$
\begin{aligned}
&(0, 0)^T \\
&(1, 0)^T \\
&(0, 1)^T \\
&(1, 1)^T
\end{aligned}
\tag{7.19}
$$

- *Linear inequality constraints representation*: This is the converse process where you use linear inequality constraints to define the lines, planes, or hyperplanes that form the facets of the convex (and bounded) set. The intersections among them define the vertices. This is precisely what is shown in Fig. 5.1. You see that the unit square is well defined by the set of 4 linear inequality constraints as in Eq. 5.31

---

[5]In the case of unbounded polygons, polyhedra, and polytopes you also need to specify the direction of the rays that emanate from some of the vertices.

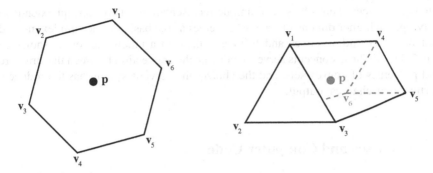

Fig. 7.10 Two convex sets. In both cases, any point **p** belonging to the convex set can be found by a linear combination of its vertices $\mathbf{v}_i$. Note that the vertices $\mathbf{v}_i$ are not the generators in Eq. 7.2 or Fig. 7.2, but rather the actual extreme points of a convex set S

$$
\begin{aligned}
x_1 &\leq 1 \\
x_2 &\leq 1 \\
x_1 &\geq 0 \\
x_2 &\geq 0
\end{aligned}
\tag{7.20}
$$

Moreover, Fig. 5.2 shows a critical departure from a Minkowski sum. This is done by adding another inequality constraint that truncates the unit square (or cube or N-cube) to create an irregular polygon, polyhedron or polytope. Generally, the square is a zonotope that can be generated by the Minkowski sum of its generators, the unit vectors $(1, 0)^T$ and $(0, 1)^T$. And the feasible sets in Figs. 7.3 and 7.5 are created in a similar way. But the truncated square in Fig. 5.2 is not necessarily a zonotope, but rather a convex polygon that can be represented by its vertices or a set of linear inequality constraints. The distinction that all zonotopes are bounded polygons, but not all bounded polygons are zonotopes will become important in the next chapter.

Regardless of whether you use the vertex or linear inequality constraint representation, what makes a set convex (as opposed to concave) is the fact that any point belonging to the set can be calculated as some linear combination of its vertices [13]. Take Fig. 7.10, where any point **p** in the convex set S can be found as a—not necessarily unique—linear combination of the 6 vertices of each of the convex sets in Fig. 7.10.

$$
x_1\mathbf{v}_1 + x_2\mathbf{v}_2 + \cdots + x_6\mathbf{v}_6
\tag{7.21}
$$

Importantly, a key concept that will be treated in detail later on is whether Eq. 7.21 is unique for a given internal point $\mathbf{p} \in S$ or not. It is clear that, for a polygon as in Fig. 7.10, an internal point does not have a unique representation (i.e., there is redundancy), but a point on the edges (or a vertex itself) has a unique representation as in Figs. 7.6 and 7.8. But for a polyhedron, a point on a face (which is a polygon by definition) does not have a unique representation (why?). Only points along the edges or vertices themselves have unique representations. This concept extends to polytopes in higher dimensions and even serves as the basis for the simplex method that relies on finding edges and following them to a unique vertex as shown in Fig. 5.3 [13]. These concepts serve to motivate the connection between the structure and properties of convex sets, and the choices the nervous system has to produce a particular mechanical output.

7.7 Exercises and Computer Code

Exercises and computer code for this section in various languages can be found at http://extras.springer.com or found by searching the World Wide Web by title and author.

References

1. F.J. Valero-Cuevas, F.E. Zajac, C.G. Burgar, Large index-fingertip forces are produced by subject-independent patterns of muscle excitation. J. Biomech. **31**, 693–703 (1998)
2. F.J. Valero-Cuevas, A mathematical approach to the mechanical capabilities of limbs and fingers. Adv. Exp. Med. Biol. **629**, 619–633 (2009)
3. E.Y. Chao, K.N. An, Graphical interpretation of the solution to the redundant problem in biomechanics. J. Biomech. Eng. **100**, 159–167 (1978)
4. C.W. Spoor, Balancing a force on the fingertip of a two-dimensional finger model without intrinsic muscles. J. Biomech. **16**(7), 497–504 (1983)
5. P.W. Brand, A. Hollister, *Clinical Mechanics of the Hand* (Mosby Year Book, St. Louis, 1993)
6. A.D. Kuo, F.E. Zajac, Human standing posture: multi-joint movement strategies based on biomechanical constraints. Prog. Brain Res. **97**, 349–358 (1993)
7. R.M. Murray, Z. Li, S.S. Sastry, *A Mathematical Introduction to Robotic Manipulation* (CRC, Boca Raton, 1994)
8. G. Strang, *Introduction to Linear Algebra* (Wellesley Cambridge Press, Wellesley, 2003)
9. Wikipedia contributors, Polytope. Wikipedia, The Free Encyclopedia. http://en.wikipedia.org/wiki/Polytope. Accessed 2 June 2015
10. Wikipedia contributors, Cartesian coordinate system. Wikipedia, The Free Encyclopedia. http://en.wikipedia.org/wiki/Minkowski_addition. Accessed 12 Feb 2015
11. Wikipedia contributors, Cartesian coordinate system. Wikipedia, The Free Encyclopedia. http://en.wikipedia.org/wiki/Zonohedron. Accessed 12 Feb 2015
12. D. Eppstein, Zonohedra and zonotopes. Math. Edu. Res. **5**, 15–21 (1996)
13. V. Chvatal, *Linear Programming* (W.H. Freeman and Company, New York, 1983)
14. F.J. Valero-Cuevas, An integrative approach to the biomechanical function and neuromuscular control of the fingers. J. Biomech. **38**, 673–684 (2005)
15. J. Dattorro, *Convex Optimization and Euclidean Distance Geometry* (Meboo Publishing, Palo Alto, 2005)
16. F.J. Valero-Cuevas, Predictive modulation of muscle coordination pattern magnitude scales fingertip force magnitude over the voluntary range. J. Neurophysiol. **83**, 1469–1479 (2000)

17. F.J. Valero-Cuevas, J.D. Towles, V.R. Hentz, Quantification of fingertip force reduction in the forefinger following simulated paralysis of extensor and intrinsic muscles. J. Biomech. **33**, 1601–1609 (2000)

18. F.J. Valero-Cuevas, M.E. Johanson, J.D. Towles, Towards a realistic biomechanical model of the thumb: the choice of kinematic description may be more critical than the solution method or the variability/uncertainty of musculoskeletal parameters. J. Biomech. **36**, 1019–1030 (2003)

19. J.M. Inouye, F.J. Valero-Cuevas, Anthropomorphic tendon-driven robotic hands can exceed human grasping capabilities following optimization. Int. J. Robot. Res. **33**, 694–705 (2013)

20. J.M. Inouye, J.J. Kutch, F.J. Valero-Cuevas, A novel synthesis of computational approaches enables optimization of grasp quality of tendon-driven hands. IEEE Trans. Robot. **28**(4), 958–966 (2012)

21. E. Moberg, Surgical treatment for absent single-hand grip and elbow extension in quadriplegia. Am. J. Bone Joint Surg. **57**, 196–206 (1975)

22. V.R. Hentz, C. Leclercq, *Surgical Rehabilitation of the Upper Limb in Tetraplegia* (WB Saunders, Philadelphia, 2002)

23. J.W. Littler, On making a thumb: one hundred years of surgical effort. J. Hand Surg. **1**(1), 35–51 (1976)

Chapter 8
Feasible Neural Commands with Mechanical Constraints

Abstract In this chapter I refine the notion of feasible neural commands by introducing the concept of functional constraints. Chapter 7 presented the geometric principles that allow us to find the structure of the sets of all feasible neural commands and feasible mechanical outputs as a function of the natural bounds on muscle activations, strengths of the muscles, routing of the tendons, and mechanics of the limb. From that perspective it is clear that producing maximal mechanical output—shown for the case of static force production—can only be achieved by a unique muscle activation pattern (i.e., there is *no* muscle redundancy). But producing sub-maximal mechanical outputs can be done by multiple muscle activation patterns (i.e., there *is* muscle redundancy). This chapter explores the nature of those multiple muscle activation patterns, and the relationships among them. I emphasize that considering the properties of the limb *plus* the functional constraints of the task (which can be mechanical, metabolic, physiological, etc.) allows us to define and find families of valid and related solutions—instead of unique solutions in isolation. These concepts continue to challenge the classical notion of muscle redundancy but, most importantly, provide perspective and computational tools to explore mechanisms by which the nervous system controls the limbs for specific tasks. That is, muscle redundancy is a function of both the limb and the task. This is directly relevant to the central questions of motor control such as optimization, learning, adaptation, dimensionality reduction, synergistic control, etc. More generally, these concepts build a case to argue that exploring and exploiting feasible activation sets is likely a more biologically tenable way in which the nervous system operates in the context of muscle redundancy and multifaceted real-world tasks.

8.1 Finding Unique Optimal Solutions Versus Finding Families of Valid Solutions

As mentioned in Chap. 1, today's dominant perspective in the study of neuromechanical systems is that muscle redundancy for net joint torque production is the central problem of motor control. This has cast the 'problem' the brain solves as

© Springer-Verlag London 2016   113
F.J. Valero-Cuevas, *Fundamentals of Neuromechanics*,
Biosystems & Biorobotics 8, DOI 10.1007/978-1-4471-6747-1_8

one where neural computation selects a specific neural command from the infinite options available—it is after all an underdetermined problem. The preceding chapters have provided the perspective and language to understand that this concept of neural optimization is founded on the notion that there are multiple options (i.e., the feasible activation set, Sect. 7.3) from which the nervous systems could select a best unique solution as per the cost function being endorsed.

How is this neural computation done? As mentioned in Chap. 5, biomechanists and neuroscientists often approach the problem of motor control as one of numerical optimization [1–6]. No one would claim that the nervous system acts like a computer running deterministic algorithms. Rather, optimization has and should be used as a metaphor [3, 4], but one that should not be taken too literally [7, 8].

The principal elements we use in algorithmic optimization all seem to be present in the neuromechanical coordination of vertebrates: the nervous system uses iterative learning and adaptation strategies to compare physiological optimization criteria (i.e., a cost function based on task performance metrics, metabolic cost, tissue stress, muscle force, intensity of α-motoneuron drive, etc.) [9]. Comparing among alternative valid neural commands drives the search for improvement until a local optimum is found and adopted. That local optimum is a local minimum or maximum of the cost function. It takes the form of an optimal activation pattern a^*, as discussed in Sect. 5.3. Importantly, such strategies employ what is called a *line search strategy* in the solution space that first finds a direction in which the cost function decreases most (or least), and then computes a step size that determines how far to move along that search direction.[1]

You can think of line search optimization methods as being blind to the global layout of the solution space. Only the value of the cost function and its local gradient (or the total cost along a particular trajectory) are used to traverse the solution space in search of an optimal solution. Allow me to make the following analogy about numerical optimization. Imagine your friends put you in a dark unfamiliar room (the solution space) and ask you to find the southernmost part of the room (the *cost function*)—but armed only with a compass and a box of short matches to make local *evaluations of the cost function*. The place where your friends put you in the room is your *feasible starting point*. What you do then is iteratively light a match, briefly see which way south is (*calculate the gradient of the cost function*) and get a glimpse of the immediate surroundings around your feet just as the match goes out. You then take a step in the dark, the size of which depends on how confident you are that you will not trip or hit a wall. After multiple iterations you will undoubtedly find a wall (an *operating constraint* at the *boundary of the solution space*), and follow that wall in a southward direction until you stop at a location in the room you consider to be your best guess of the solution (the *local minimum*), and stop. That location is your *optimal solution* because movement in any direction would either increase your cost

---

[1]Stochastic methods like simulated annealing [10] or Markov chains [11] are sometimes used instead of true gradient descent methods. But in principle these methods are a way to overcome computational difficulties when calculating best search directions in complex cost landscapes, or exploring high-dimensional solution spaces where calculating an exact gradient is too computationally expensive or slow.

or is not possible because of a wall. Although this analogy applies in general, there are many distinct types of optimization (see [5, 12, 13]). From this perspective it is clear to see how important the choice of cost function, step size, and efficiency of each iteration are to the speed and quality of your results—especially for nonlinear systems or cost functions, and for high-dimensional solutions spaces. It is also clear why there can be important and heated debates about these choices, and how best practices for one problem may not apply to another.

Finding feasible sets using using computational geometry methods [14] is an alternative to the line search approach—in fact it is its mathematical *dual* or counterpart. To follow the analogy above, finding feasible sets (more precisely, finding convex polytopes using vertex enumeration methods [15]) is like turning on the lights in the previously dark room. You can now see every wall, light switch, exit, and obstacle. In a word, it allows us to see and interpret the solution space and find *families* of solutions for a variety of cost functions. For example, it allows you to see all locations on the southernmost wall. Thus, the choice of cost function is no longer as critical. Clearly, a cost function will still allow you to select a particular solution, but you can be more flexible as you now work with *families* of solutions that are generally valid (i.e., good enough or sub-optimal) instead of insisting on the one unique solution you found, which may not be global, generalizable, or robust.

To be fair, line search optimization algorithms can be extremely efficient, and operate well in high-dimensional problems. Thus the reasons for their development, selection, and use must not be underestimated. By comparison, computational geometry methods are naturally much more computationally costly. Thankfully these methods are now developed to the point that applying them to relatively high-dimensional tendon-driven limbs is tractable [11, 16–19].

This chapter compares and contrasts line search and computational geometry approaches with the goal of showing how they are complementary. A brief introduction to line search methods in the context of tendon-driven limbs was given in Sect. 5.3. Here I point out how computational geometry approaches can provide unique insights into understanding the neural control of tendon-driven limbs. Calculating, describing, and interpreting feasible sets is a particularly powerful approach to neuromechanical analysis. It is a complement and alternative to the dominant emphasis on finding individual optimal solutions. But more importantly, it builds foundation for a case, which argues that exploring and exploiting feasible sets in the activation, muscle force, joint torque, or output wrench spaces is a biologically plausible way in which the nervous system may learn and produce the multiple, multifaceted tasks of everyday life in the context of muscle redundancy.

8.2 Calculating Feasible Sets for Tasks with Functional Constraints

As mentioned above, the feasible sets we are exploring are convex polytopes in their particular multidimensional spaces (e.g., activation, muscle force, joint torque, or output wrench spaces) as shown in Fig. 7.3. These polytopes describe and contain all feasible vectors of $\mathbf{a}$, $\mathbf{f}_m$, $\boldsymbol{\tau}$, and $\mathbf{w}$, respectively.

The simplest case is to define the set of feasible activations as the unit positive N-cube, and map it into the feasible muscle force, torque, and wrench sets as shown in Chap. 7. This gives us a global view of the raw mechanical capabilities of the limb without any additional constraints. From this, we can find the maximal possible output wrench at the endpoint *in every endpoint force and torque direction*. Figure 7.7 shows the maximal *endpoint force* for two sample directions. To find each of these cases using linear programming, Sect. 5.3 shows you need to define a cost function and constraints to define the desired direction, and then use a numerical algorithm to maximize force magnitude in that given direction. However, if the feasible wrench set is known (the feasible force set in Fig. 7.7 for this example), all we need to do is to send a ray from the origin in any desired direction. The point where that ray meets the boundary of the feasible force set is, by definition, the maximal possible force in that direction [20].

More general cases, however, do not have the generic unit positive N-cube as their feasible activation set. Take, for example, the linear programming problem in Fig. 5.2, where we have the system of inequality constraints that define the unit square, plus the constraint $x_1 + x_2 \le c$ as shown in Eq. 5.31. The feasible activation set is thus a convex polygon that is a subset of the unit square, Fig. 5.2.

This same principle applies in higher dimensions. Consider the questions I approached for my doctoral dissertation: what are the feasible activation and feasible force sets for a human index finger [20, 21]? In this case we have an extended version of Eq. 5.22 where we have a 3D finger with 4 DOFs and 7 muscles, shown in Fig. 8.1 and Eq. 8.1

$$\mathbf{w} = \begin{pmatrix} f_x \\ f_y \\ f_z \\ \tau_z \end{pmatrix} = J^{-T} R F_0 \begin{pmatrix} a_1 \\ a_2 \\ a_3 \\ a_4 \\ a_5 \\ a_6 \\ a_7 \end{pmatrix} = J^{-T} R F_0 \, \mathbf{a}$$

$$\mathbf{w} \in \mathbb{R}^4 \qquad\qquad (8.1)$$
$$J^{-T} \in \mathbb{R}^{4 \times 4}$$
$$R \in \mathbb{R}^{4 \times 7}$$
$$F_0 \in \mathbb{R}^{7 \times 7}$$
$$\mathbf{a} \in \mathbb{R}^7$$

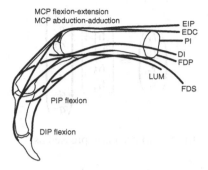

Fig. 8.1 Index finger model. This model has 4 kinematic DOFs (metacarpophalangeal flexion-extension, metacarpophalangeal abduction-adduction, proximal interphalangeal flexion, and distal interphalangeal flexion), and therefore a 4D output wrench (3 forces and one torque perpendicular to the plane of finger flexion-extension). It is driven by 7 muscles: *flexor digitorum profundus* (FDP), *flexor digitorum superficialis* (FDS), *extensor indices propius* (EIP), *extensor digitorum comunis* (EDC), *lumbrical* (LUM), *first dorsal interosseous* (DI) and *first palmar interosseous* (PI)

You could of course map the 7D N-cube as the feasible activation set, but what you would get is a 4D convex polytope in wrench space (the feasible wrench set) with 3 dimensions of force and 1 dimension of torque. While a valid object, that polytope is not possible to show in 3D with meaningful axes. Moreover the presence of a non-zero torque element in the feasible wrench set poses problems from the functional perspective. That is, producing a force against a surface with a so-called *point contact* with friction like the fingerpad [22] while there is an output torque τ_z at the fingertip would cause the finger to rotate and lose its posture. Therefore, it is of greater functional interest to know the capabilities of the finger as it produces a static force against a surface with the fingertip while the fingertip does not rotate about the contact point. These constraints are discussed in detail in [20, 23] but, briefly, this means that you want to find the feasible output while the torque in the output wrench τ_z is constrained to be zero.

How does one do that? Consider Eq. 8.1, where the fourth row describes how muscle activations contribute to the production of the τ_z element of the output wrench **w**. This same concept was first described in Sect. 5.3 when demonstrating the use of linear programming to neuromuscular problems. There we showed how Eq. 8.1 can be written by defining the matrix H

$$H = J^{-T} R F_0 \tag{8.2}$$

with row vectors $\mathbf{h}_i^T$ as follows

$$\begin{pmatrix} f_x \\ f_y \\ f_z \\ \tau_z \end{pmatrix} = \begin{bmatrix} \mathbf{h}_1^T \\ \mathbf{h}_2^T \\ \mathbf{h}_3^T \\ \mathbf{h}_4^T \end{bmatrix} \begin{pmatrix} a_1 \\ a_2 \\ a_3 \\ a_4 \\ a_5 \\ a_6 \\ a_7 \end{pmatrix} \tag{8.3}$$

Here, the constraint of setting τ_z to zero is expressed as

$$\tau_z = \mathbf{h}_4^T \cdot \begin{pmatrix} a_1 \\ a_2 \\ a_3 \\ a_4 \\ a_5 \\ a_6 \\ a_7 \end{pmatrix} = 0 \tag{8.4}$$

Therefore the feasible activation set is the portion of the 7D N-cube that intersects the plane defined by the linear equation $\mathbf{h}_4^T \cdot \mathbf{a} = 0$. All activation patterns contained in this 6D subset of the 7D space will meet the constraint of having no output τ_z. This will create the 3D feasible force set for the index finger while it produces no endpoint torques.[2] As an explanation of this, consider Fig. 8.2 as a direct analogy. Here you have the intersection of a unit cube (3D object) with a 2D plane (i.e., one equation in 3 variables). Their intersection produces a feasible set that is a 2D plane embedded in 3D space because the dimensionality of the 3D object was reduced by 1 (i.e, the introduction of 1 constraint typically reduces the feasible space by 1 dimension). All points that are both on the plane *and* in the unit cube will satisfy all constraints.

To make the force production task even more realistic and functional, consider that you are interested in all activation patterns that not only produce no output τ_z, but also produce no side-to-side fingertip forces. These are the kinds of forces you would use to roll a pencil with the index fingertip: any lateral forces would make the pencil twist and fall from your grasp. In this example, such side-to-side forces are in the z direction. Therefore, you would want to enforce 2 constraints

$$\begin{pmatrix} f_z \\ \tau_z \end{pmatrix} = \begin{bmatrix} \mathbf{h}_3^T \\ \mathbf{h}_4^T \end{bmatrix} \begin{pmatrix} a_1 \\ a_2 \\ a_3 \\ a_4 \\ a_5 \\ a_6 \\ a_7 \end{pmatrix} = \begin{pmatrix} 0 \\ 0 \end{pmatrix} \tag{8.5}$$

---

[2]If you are curious about what this looks like see Fig. 8.6.

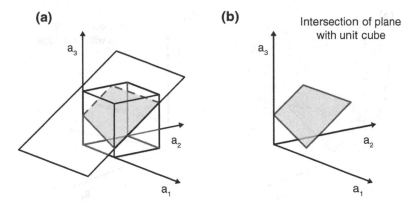

Fig. 8.2 **a** Feasible activation set for 1 functional constraint. **b** The intersection of a 3D unit cube with a 2D plane is a 2D plane embedded in 3D

Using the analogy in Fig. 8.2 again, the intersection of the cube with 2 planes—if it exists—is a 1D line. That is, the activation patterns that meet all constraints lie on a 1D line embedded in 3D space. See Fig. 8.3.

Turning back to our 7-muscle finger, the feasible activation set for all possible forces in the plane of finger flexion must meet both $\mathbf{h}_4^T \cdot \mathbf{a} = 0$ and $\mathbf{h}_3^T \cdot \mathbf{a} = 0$. That feasible activation set is a 5D convex set ($7 - 2 = 5$) embedded in 7D. Figure 8.4 shows the projection of the feasible activation set onto the feasible force set, which is naturally 2D because we are only plotting the 2 output dimensions f_x and f_y that are not constrained to be 0.

Note the importance of the fact that every task constraint that is added reduces the dimensionality of the feasible activation set by 1. And when the dimensionality of the feasible activation set is 1 (i.e., a point), then there is only a unique solution. Thus a limb with 7 muscles can only meet 6 functional constraints simultaneously: the direction of the force (constraints in f_x, f_y, and f_z force) are 3, the value of the associated endpoint torques (constraint in τ_z) is another, and the properties of the stiffness of the limb (orientation and shape of the stiffness ellipsoid) would be 2 more [1, 18, 20, 24–26].

8.3 Vertex Enumeration in Practice

How does one find the intersection of a 7D unit cube with 2 hyperplanes in 7D space? This is the *vertex enumeration problem*, defined as finding the vertices of the polytope that meets those constraints. It is important to note that the problem of finding convex polytopes from linear inequality constraints is a part of a rich area of mathematics called computational geometry [14, 27]. Therefore, I only touch

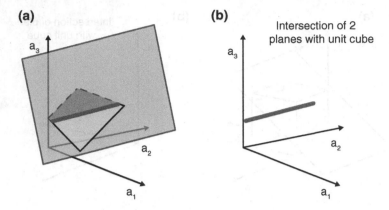

Fig. 8.3 **a** Feasible activation set for 2 functional constraints. **b** The intersection of a 3D unit cube with two 2D planes, if it exists, is a 1D line embedded in 3D

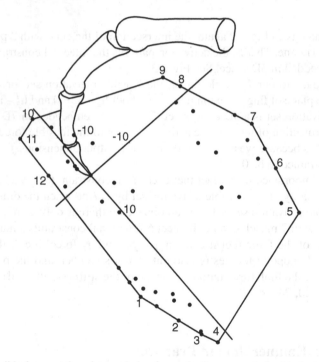

Fig. 8.4 Feasible force set for a human index finger, for forces constrained to be in the plane of finger flexion from [20, 21]. That is, these fingertip forces are not accompanied by side-to-side force components f_z, or torque at the fingertip τ_z. This convex polygon is found by first finding the feasible activation set that is a 5D polytope embedded in 7D space, which is defined by 46 vertices [20]. Mapping all 46 vertices of the feasible activation set produce the feasible force set. In this case, 12 vertices become extreme points and define the convex hull. There are 34 internal points that are the projection of the vertices of the feasible activation set that do not become extreme points in this 2D output space. See Figs. 7.6 and 8.5. Figure adapted from [21]

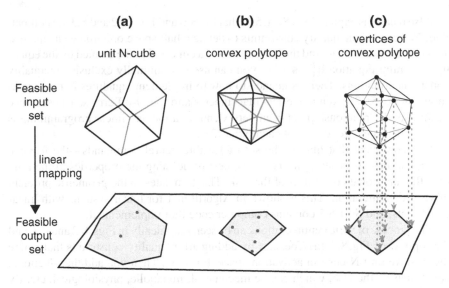

Fig. 8.5 *Top row* Schematic representation of Algorithm 1. Note that the addition of inequality constraints reduces the size and distorts the shape of the original *feasible input set*—the positive unit N-cube—and transforms it into a new convex polytope. The more faces the convex polytope has the more vertices are required to define it. Columns **a–c**: Schematic representation of Algorithm 2. As illustrated in Fig. 7.6, the projection of the feasible input set produces a *feasible output set* that is a convex hull that defines the feasible muscle force, joint torque, wrench, or output force sets, depending on the framing of the problem. Figure 8.1 shows a real-world example of mapping a feasible activation set in $\mathbb{R}^7$ onto the feasible force set in the plane of finger flexion in $\mathbb{R}^2$. Note that the vertices that become internal points of the feasible output set hint at the shape of the feasible input set, which is hard to describe or visualize given its high-dimensionality. This is an area of active research [18]

upon the most necessary points, and the reader is encouraged to see the references mentioned.

Recall that in Sect. 7.6 we introduced two types of representations of a convex polytope. In this case, we are interested in finding the vertices of a high-dimensional convex polytope defined by a set of linear inequality constraints [15] as in

$$A\mathbf{x} \leq \mathbf{b} \tag{8.6}$$

where $A \in \mathbb{R}^{M \times N}$ is a matrix containing M equations in N variables, $\mathbf{x} \in \mathbb{R}^N$, and $\mathbf{b} \in \mathbb{R}^M$ is a vector of constants.

Do you notice the similarity to the phrasing of the canonical form of the linear programming problem in Eq. 5.9? Notably, however, there is no need to have a cost function. That is, we are not interested in finding the maximal force *in a particular direction* in the plane of finger flexion as in Eq. 5.22. Instead, we are interested in finding *all possible maximal forces* in the plane of finger flexion.

Also recall, as explained in Sect. 5.3, that Eq. 8.6 and Figs. 8.2 and 8.3 are compatible. You can use inequality constraints to define a half space or enforce an equality. Thus, when you want to find the intersection of a cube and plane defined by the equality constraint equation $\mathbf{h}_4^T \cdot \mathbf{a} = 0$, you can use two mutually exclusive inequality constraint equations. There is nuance to how to implement equalities in practice as shown in Eq. 5.22, given numerical inaccuracies and round-off errors. You see once again why the relationship between vertex enumeration and linear programming is so close that they are 'duals' of each other [28].

It is now clear that finding the set of feasible neural commands—the feasible activation set—is the straightforward process of defining the properties of the limb *plus* the functional constraints of the task. This translates to the geometric problem of vertex enumeration. This is shown in Algorithm 1 for linear systems with linear constraints, and Sect. 8.7 contains computer code that implements it.

The process of vertex enumeration can be seen graphically in Figs. 8.2 and 8.3, and the top row of Fig. 8.5. (a) You start by adding all inequality constraints that define the positive unit N-cube in activation space, Fig. 8.5a. Then you add the functional constraints of the task, which can be mechanical, metabolic, physiological, etc. By doing this the N-cube is transformed into a convex polytope reduced in size (why?) and distorted in shape (why?) that is often of lower-dimension (why?) (b). The new convex polytope can be fully described by its vertices (c), or faces (linear inequality constraint equations) that compose it.

Algorithm 1 Finding the vertices of a feasible activation set

---

Require: Linear system with linear equality or inequality constraints
 $\rightarrow$ Find $H = [J^{-T} \ R \ F_0]$
 $\rightarrow$ Define the linear constraint equations that define your task, either as rows of H or otherwise
 $\rightarrow$ Define activation space as the unit N-cube $0 \le \mathbf{a} \le 1$
 $\rightarrow$ Assemble $A\mathbf{a} \le \mathbf{b}$
 $\rightarrow$ Use $A\mathbf{a} \le \mathbf{b}$ to find the intersection of all inequality or equality constraints. This can be done
     with code such as cdd by Avis and Fukuda [15]
 $\rightarrow$ The intersection of all inequality or equality constraints, if it exists, forms a convex polytope.
     It can be requested in vertex or linear inequality constraints representation (See Sect. 7.6)
 $\rightarrow$ This polytope is the feasible activation set [20, 21]

How does one find the feasible output set? As illustrated in Fig. 8.5 the projection of the *feasible input set* produces a *feasible output set*—a convex hull that defines the feasible muscle force, joint torque, wrench, or force sets depending on the framing of the problem. Figure 7.3 shows this sequence of mappings. Please note that, in general, the feasible input set need not be the feasible activation set. For example, Fig. 7.3 shows that you can be doing your analysis in the feasible torque set, and call that the feasible input set, to produce the feasible endpoint wrench set, which would be your feasible output set. It all depends on the analysis you are doing.

Having said this, mapping the feasible input set into the feasible output set translates into the computational geometry problem of finding the *convex hull* for a cloud of points in the space into which the mapping takes place. This is done as per Algorithm 2. An example of a convex polyhedron representing a feasible output set is shown in Fig. 8.6—the 3D feasible force set for the human index finger [20, 21].

Algorithm 2 Finding the feasible output set

Require: Vertices of the feasible input set. These can be obtained as per Algorithm 1 from the matrices F_0, R, J^{-T} or their combinations, depending on your definition of what is the input as per Fig. 7.3

→ Project every vertex of the feasible input set by the matrices of interest that produce the desired output. For example,

if you consider the *feasible activation set* as your input

then

    → Multiply all vertices by F_0 to find the *feasible muscle force set*

    → Multiply all vertices by $[R\ F_0]$ to find the *feasible joint torque set*

    → Multiply all vertices by $[J^{-T}\ R\ F_0]$ to find the *feasible endpoint wrench set* or *feasible endpoint force set* depending on the constraints you enforce

end if

return All pairs of vertices of the feasible activation space and the output they produce

→ Apply the convex hull operation to the projected vertices to find the convex polyhedron, polygon, or polytope (depending on the dimensionality of the output space). This can be done with command `convhull` in MATLAB or `qconvex` of the Qhull library found at www.qhull.org

→ The projected vertices that create the convex hull are those that produce maximal output by their linear combinations as shown in, for example, Fig. 7.8

→ That convex polyhedron, polygon, or polytope is the feasible output set [20, 21]

Fig. 8.6 3D feasible force set. This convex polyhedron shows the 3D force capabilities $(f_x, f_y, f_z)^T$ of the human index finger as calculated from data collected in cadaver fingers [29, 30]. Here the constraint that is enforced is that the fingertip should produce 3D forces without any accompanying endpoint torque τ_z. The full output wrench is the 4D $\mathbf{w} = (f_x, f_y, f_z, \tau_z)^T$, but enforcing the constraint on $\tau_z = 0$ produces the 3D feasible force set shown. Figure adapted with permission from [30]

8.4 A Definition of Versatility

The insight we gain from the graphical interpretation of the feasible actions of tendon-driven limbs allows us to ask several fundamental questions about motor control. For example, what properties should the feasible activation or output sets have to be *adequate*, *desirable*, or even *optimal*? What anatomical or neural features produce those functional properties? The discussion in Sect. 7.5 indicates that such questions cannot be answered definitively.

> A definition of *versatility* is that the endpoint of the limb be able to produce static force in every possible direction at every posture. That is, that feasible output set must include the origin—be it in the joint torque, endpoint wrench, or endpoint force spaces [31].

This concept of versatility is very much related to the concept of *force closure* for grasp analysis [22, 32]. That is, if a system is able to resist a perturbation coming from any direction by using its actuators, then the object will remain in the grasp. Thus we see that a tendon-driven limb can be considered versatile in the sense that it can resist (or produce) a wrench at the endpoint in any direction only if the origin is included in the set of feasible wrenches. This definition is closely tied to the concept of *controllability*, but not exactly. Mathematically, output controllability is defined as the ability to reach any state, from any other state, in finite time [33]. Roboticists go on to quantify how uniformly versatile a robotic limb is by finding the radius of the largest ball, centered on the origin, that can fit in the feasible output space—which necessarily means that the feasible output set must enter some portion of every octant [16, 34, 35].

This concept of versatility can be cast at the level of actuation of all DOFs: that at any time one can span every dimension of the space in both positive and negative joint torque directions. That is, in the torque space, versatility can be defined as being able to produce both positive and negative torques at each joint. In endpoint force space, versatility would be the ability to produce positive and negative forces in the x, y, and z directions, etc.

Figure 7.5 shows a case where having only muscles 1 through 3 does not suffice for the limb to be versatile because their feasible net joint torque set does not include positive torques at the second joint. It is only when muscle 5 is added that versatility is achieved. By contrast, Fig. 7.9 shows the case of 8 generators, where versatility was already achieved with generators 1 through 3—independently of what 'space' we are in.

Last, in the case of fingertip endpoint forces in Fig. 8.6, the origin of force production is at the fingertip, which is inside the feasible force set. There we see that the fingertip could feasibly produce or resist forces in every 3D direction: toward an object being grasped, or sideways to rotate the object, or away from the object. In [30] we show how the loss of some muscles can easily remove versatility. While it is trivial to say that paralysis weakens, being able to show exactly how and in which directions the weakening occurs is critical to understand the disability—and to design therapeutic or surgical strategies that restore as much versatility as possible [30].

How does one achieve versatility in tendon-driven limbs? The argument about versatility is often made in the endpoint wrench space where force closure is defined. It makes sense to evaluate the final, most visible mechanical output of the system. However, it is not entirely obvious what properties of the system lead to versatility in the endpoint wrench space because it is the cumulative product of all of the transformations in Fig. 7.3. There are several properties of convex sets that help us establish versatility in a more anatomically intuitive way.

As described in [31], it is possible to show that if the feasible joint torque set includes the origin, then the feasible wrench set will also include the origin, and vice versa. Consider Figs. 7.3 and 7.4, which show that the system is versatile because the feasible joint torque and feasible wrench sets both include the origin. The following argument proves this:

- Feasible sets are convex polytopes.
- The vertices of a convex polytope define its size and shape.
- Convex sets remain convex under linear mapping [15, 28]. Matrix multiplication is an example of linear mapping as is shown in Figs. 7.3, 7.6, and 8.5.
- Under linear mapping, the extreme points (vertices and boundaries) of the output convex set are produced by the extreme points of the input convex set.
- Another way to say this is that internal points of the input convex set can never become extreme points in the output convex set.
- Therefore, if the zero point (i.e., the origin) is an internal point of the feasible joint torque set, then the origin will be an internal point of the feasible wrench set. Thus a system will be versatile in wrench space (i.e., the feasible wrench space will include the origin) if the feasible joint torque set includes the origin in joint torque space.

8.5 How Many Muscles Should Limbs Have to be Versatile?

Having a working definition of versatility that can be addressed at the level of the 'intermediate' joint torque space (as shown in Fig. 7.3) has multiple advantages. Most importantly, it is intuitive because the details of the generators that produce the feasible joint torque set come directly from the anatomical routing of the tendons, as shown in Fig. 7.5. Any constraint equations that are added in activation space that modify the initial N-cube can also be readily interpreted at the level of individual muscle actions in torque space. And last, the interactions between neural and anatomical constraints in health and disease can be interpreted at a very intuitive level in torque space.

Seen from this perspective, the question of how many muscles are needed for a versatile limb should be: what is the minimal number of well-routed tendons required

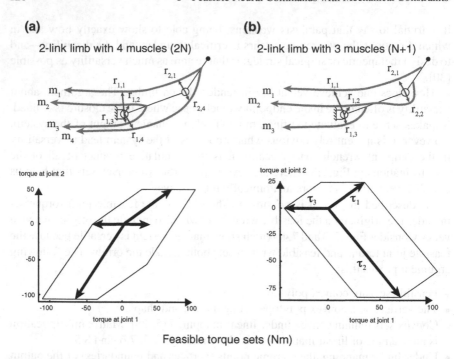

Fig. 8.7 Feasible joint torque sets for systems with different numbers of muscles. N is the number of kinematic DOFs, in this case 2 for a planar 2 joint system. **a** Roboticists often prefer to have 2 tendons per DOF, or $2N$ tendons, to achieve versatility [36]. **b** But the minimal number of muscles is in fact $N + 1$, so long as they are appropriately routed. Figure adapted with permission from [31]

to produce a feasible torque set that includes the origin? Considering that muscles can only pull actively, one would intuitively think that you need 2 muscles per DOF (1 tendon on each side of the joint)—a so-called 2-N design in robotics where N would be the number of kinematic DOFs.[3] But Fig. 8.7 shows that you really only need 1 more muscle than the number of kinematic DOFs—an $N + 1$ design—provided that its tendons are *well-routed*.

A graphical way to interpret the minimal number of muscles is that, as shown in Fig. 8.7, there are enough vectors such that their Minkowski sum includes the origin. As shown in Fig. 7.5b, this means that these vectors need to be properly directed—and the moment arms at each joint define their direction:

- If a tendon crosses only 1 DOF, its vector in torque space is along that joint torque dimension.
- If a tendon crosses more DOFs, its direction is defined by the ratio of moment arms at each joint.

---

[3]Recall that robotic designs are traditionally dominated by torque-driven limbs, hence the letter N stands for number of DOFs. In this book, as seen in Part I, I use N as the number of muscles. This unfortunate reuse of a given variable is not uncommon in mathematical descriptions.

Therefore, there is a direct and intuitive relationship between the anatomical routing of tendons and the versatility of the system.

Even though versatility is a necessary condition for a useful limb, it is not sufficient. As mentioned in Sect. 7.5, it is not sufficient for the feasible output set to simply include the origin. The size and shape of the polytope (i.e., the extent to which it spans the different octants) matters—and every muscle can potentially contribute to these properties as shown in Figs. 7.5 and 7.9.

There is an example tendon routing that is particularly interesting that has to do with the anatomy of human fingers. Consider Fig. 7.5 once again, and visualize the feasible joint torque set when muscles 1, 2, 3, and 5 are present. Such a finger would be versatile, but we see intuitively that quadrant IV is spanned only minimally. Therefore, the addition of a generator like muscle 4 could be considered advantageous.

Now consider that generator from an anatomical perspective: increasing the reach of the zonotope in quadrant IV requires a muscle with a positive moment arm at the first joint (i.e., be a flexor of the first joint), and a negative moment arm at the second joint (i.e., be an extensor of the second joint), as per the sign conventions of moments in Chap. 4. The actual vector direction depends on the ratio of moment arms. This is unusual because most tendons tend to be routed on the same side of the limb, like the long flexors or extensors of the fingers that flex or extend, respectively, all joints that they cross. Interestingly, such mathematically reasonable, but anatomically odd, *cross-over* tendon routings exist in the human fingers. They are the tendinous apparatus often called the *extensor mechanism* that flexes the metacarpophalangeal joint, but extends the proximal and distal interphalangeal joints [37, 38]. Therefore, this may be an evolutionary adaptation trait that provides versatility of finger function given its anatomical constraints (i.e., not having dedicated finger muscles that only act on the middle joints of the fingers). This specific adaptation in the fingers may be a response to the anatomical limitation that all muscles of the fingers are actually located in the palm of the hand or forearm, and always proximal to the knuckles (the metacarpophalangeal joints). Thus, fingers are controlled at a distance like marionettes. This is not the case in the larger limbs where there are muscles dedicated to act on individual joints, like the elbow, wrist, knee, and ankle.

8.6 Limb Versatility Versus Muscle Redundancy

These concepts also enable us to revisit the common notion that vertebrates have *too many muscles* or that muscle redundancy is the central problem of motor control. In a sense the question of redundancy ('How many muscles is too many?') is the counterpart to the question of 'What is the minimal number of muscles?' On one hand, the concept of versatility as discussed in Sect. 8.4 begs the question of which muscles one would want to give up. Specifically in Figs. 7.5 and 7.9, the question of deciding which regions of which octants of the feasible output space should be favored or discouraged translates into how many muscles one should have and how they should be routed. Even for the minimal case of N+1 muscles, we must contend with the asymmetry of tendon actuation (i.e., muscles can only actively pull on tendons). Thus one must have more muscles than kinematic DOFs. Producing sub-maximal output is always an underdetermined problem, as shown in Fig. 8.8.

Therefore, having more (or even many more) muscles than kinematic DOFs is in reality an appropriate anatomical adaptation for versatility in tendon-driven limbs. The fact that each muscle affects the size and shape of the feasible output sets implies that having many more muscles than kinematic DOFs is actually functionally desirable despite making the selection of individual muscle activation patterns underdetermined.

This begs the question of when a problem is *too* underdetermined. This is a key question that is invariably anthropomorphized in our literature because our mathematical and computational tools to solve underdetermined systems have specific formulations and requirements as discussed in Chap. 5. We will come back to this point, but for now let us consider Algorithm 1, shown graphically in Figs. 8.2, 8.3, and 8.5. It is clear from the material in this chapter that:

- The feasible activation set contains all possible solutions to a problem, and is the set from which any solution must come regardless of how it is found (be it by line-search optimization, simulated annealing, random sampling, memory, trial-and-error, etc.)
- The size and shape of the feasible activation set depends on the interactions among the number of muscles and the number and type of functional constraints

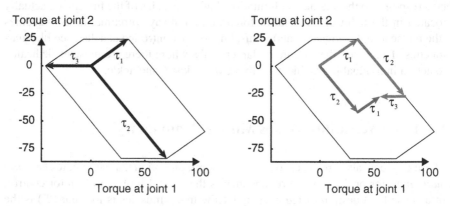

Fig. 8.8 Versatile tendon-driven limbs must also be redundant. There are no unique muscle coordination patterns to produce sub-maximal static forces—even for the minimal number of muscles needed to make the limb versatile, $N+1$. This implies that reaching any internal point in the feasible joint torque set (and in the feasible wrench set) can be achieved by multiple solutions (i.e., vectors in the feasible activation space). Thus versatility implies muscle redundancy. Figure adapted with permission from [31]

- The feasible activation set will necessarily reduce in dimensionality and/or shrink in size as constraints are added
- Therefore, having many muscles provides the capability of being able to satisfy more functional constraints

Having more (or even many more) muscles than kinematic DOFs is thus an appropriate anatomical adaptation to satisfy multiple constraints. *'Real world tasks'* are the subject of *neuroethology*, which includes the evolutionary and comparative study of the mechanistic control of natural behavior by the nervous system [39–41]. The link I highlight here is that natural behavior is defined by multiple and often competing constraints, which would naturally reduce the feasible activation set (and therefore the feasible output sets) much more than the reductionist experimental tasks we often study [20, 24, 42]. Therefore, the extent and quality of redundancy cannot be expressed simply as the difference between the number of muscles and the number of kinematic DOFs. It is the structure and dimensionality of the feasible activation set that helps us see muscle redundancy from a neuroethological perspective.

8.7 Exercises and Computer Code

Exercises and computer code for this chapter in various languages can be found at http://extras.springer.com or found by searching the World Wide Web by title and author.

References

1. E.Y. Chao, K.N. An, Graphical interpretation of the solution to the redundant problem in biomechanics. J. Biomech. Eng. **100**, 159–167 (1978)
2. R.D. Crowninshield, R.A. Brand, A physiologically based criterion of muscle force prediction in locomotion. J. Biomech. **14**(11), 793–801 (1981)
3. E. Todorov, M.I. Jordan, Optimal feedback control as a theory of motor coordination. Nat. Neurosci. **5**(11), 1226–1235 (2002)
4. S.H. Scott, Optimal feedback control and the neural basis of volitional motor control. Nat. Rev. Neurosci. **5**(7), 532–546 (2004)
5. R. Shadmehr, S. Mussa-Ivaldi, *Biological Learning and Control: How the Brain Builds Representations, Predicts Events, and Makes Decisions* (MIT Press, Cambridge, 2012)
6. F.J. Valero-Cuevas, M. Venkadesan, E. Todorov, Structured variability of muscle activations supports the minimal intervention principle of motor control. J. Neurophysiol. **102**, 59–68 (2009)
7. G.E. Loeb, Optimal isn't good enough. Biol. Cybern. **106**(11–12): 757–765, (2012)
8. A. De Rugy, G.E. Loeb, T.J. Carroll, Muscle coordination is habitual rather than optimal. J. Neurosci. **32**(21), 7384–7391 (2012)

9. B.I. Prilutsky, Muscle coordination: the discussion continues. Mot. Control **4**(1), 97–116 (2000)
10. J.S. Higginson, R.R. Neptune, F.C. Anderson, Simulated parallel annealing within a neighborhood for optimization of biomechanical systems. J. Biomech. **38**(9), 1938–1942 (2005)
11. V.J. Santos, C.D. Bustamante, F.J. Valero-Cuevas, Improving the fitness of high-dimensional biomechanical models via data-driven stochastic exploration. IEEE Trans. Biomed. Eng. **56**, 552–564 (2009)
12. F.J. Valero-Cuevas, H. Hoffmann, M.U. Kurse, J.J. Kutch, E.A. Theodorou, Computational models for neuromuscular function. IEEE Rev. Biomed. Eng. **2**, 110–135 (2009)
13. P.E. Gill, W. Murray, M.H. Wright, *Practical Optimization* (Academic Press, New York, 1981)
14. M. De Berg, M. Van Kreveld, M. Overmars, O.C. Schwarzkopf, *Computational Geometry* (Springer, New York, 2008)
15. D. Avis, K. Fukuda, A pivoting algorithm for convex hulls and vertex enumeration of arrangements and polyhedra. Discret. Comput. Geometry **8**(3), 295–313 (1992)
16. J.M. Inouye, J.J. Kutch, F.J. Valero-Cuevas, A novel synthesis of computational approaches enables optimization of grasp quality of tendon-driven hands. IEEE Trans. Robot. **28**(4), 958–966 (2012)
17. J.J. Kutch, F.J. Valero-Cuevas, Challenges and new approaches to proving the existence of muscle synergies of neural origin. PLoS Comput. Biol. **8**(5), e1002434 (2012)
18. F.J. Valero-Cuevas, B.A. Cohn, H.F. Yngvason, E.L. Lawrence, Exploring the high-dimensional structure of muscle redundancy via subject-specific and generic musculoskeletal models. J. Biomech. **48**(11), 2887–2896 (2015)
19. V.J. Santos, F.J. Valero-Cuevas, A Bayesian approach to biomechanical modeling to optimize over large parameter spaces while considering anatomical variability. Conf. Proc. IEEE Eng. Med. Biol. Soc. **6**, 4626–4629 (2004)
20. F.J. Valero-Cuevas, F.E. Zajac, C.G. Burgar, Large index-fingertip forces are produced by subject-independent patterns of muscle excitation. J. Biomech. **31**, 693–703 (1998)
21. F.J. Valero-Cuevas, Muscle coordination of the human index finger. Ph.D. thesis, Stanford University, Stanford, CA (1997)
22. R.M. Murray, Z. Li, S.S. Sastry, *A Mathematical Introduction to Robotic Manipulation* (CRC, Boca Raton, 1994)
23. F.J. Valero-Cuevas, Predictive modulation of muscle coordination pattern magnitude scales fingertip force magnitude over the voluntary range. J. Neurophysiol. **83**(3), 1469–1479 (2000)
24. K.G. Keenan, V.J. Santos, M. Venkadesan, F.J. Valero-Cuevas, Maximal voluntary fingertip force production is not limited by movement speed in combined motion and force tasks. J. Neurosci. **29**, 8784–8789 (2009)
25. C.W. Spoor, Balancing a force on the fingertip of a two-dimensional finger model without intrinsic muscles. J. Biomech. **16**(7), 497–504 (1983)
26. R. Balasubramanian, Y. Matsuoka, Biological stiffness control strategies for the anatomically correct testbed (ACT) hand, in *IEEE International Conference on Robotics and Automation. ICRA 2008.* (IEEE, 2008), pp. 737–742
27. U.G. Center, QHull. UIUC Geometry Center, QHull Computational Geometry Package (2004)
28. V. Chvatal, *Linear Programming* (W.H. Freeman and Company, 1983)
29. F.J. Valero-Cuevas, J.D. Towles, V.R. Hentz, Quantification of fingertip force reduction in the forefinger following simulated paralysis of extensor and intrinsic muscles. J. Biomech. **33**, 1601–1609 (2000)
30. F.J. Valero-Cuevas, V.R. Hentz, Releasing the A3 pulley and leaving flexor superficialis intact increases pinch force following the Zancolli lasso procedures to prevent claw deformity in the intrinsic palsied finger. J. Orthop. Res. **20**, 902–909 (2002)
31. F.J. Valero-Cuevas, A mathematical approach to the mechanical capabilities of limbs and fingers. Adv. Exp. Med. Biol. **629**, 619–633 (2009)
32. B. Siciliano, O. Khatib, *Springer Handbook of Robotics* (Springer, New York, 2008)
33. K. Ogata, *Modern Control Engineering*, 5th edn. (Prentice Hall, New Jersey, 2009)
34. A.T. Miller, P.K. Allen, Examples of 3D grasp quality computations, in *IEEE International Conference on Robotics and Automation*, vol. 2. (IEEE, 1999), pp. 1240–1246

35. A.T. Miller, P.K. Allen, Graspit! A versatile simulator for robotic grasping. IEEE Robot. Autom. Mag. **11**(4), 110–122 (2004)
36. J.M. Inouye, F.J. Valero-Cuevas, Anthropomorphic tendon-driven robotic hands can exceed human grasping capabilities following optimization. Int. J. Robot. Res. (2013)
37. P.W. Brand, A. Hollister, *Clinical Mechanics of the Hand* (Mosby, Maryland Heights, 1993)
38. E. Zancolli, *Structural and Dynamic Bases of Hand Surgery*, 2nd edn. (Lippincott, Philadelphia, 1979)
39. J.M. Camhi, *Neuroethology: Nerve Cells and the Natural Behavior of Animals* (Sinauer Associates, Sunderland, 1984)
40. J.P. Ewert, *Neuroethology: An Introduction to the Neurophysiological Fundamentals of Behavior* (Springer, New York, 1980)
41. S. Giszter, V. Patil, C. Hart, Primitives, premotor drives, and pattern generation: a combined computational and neuroethological perspective. Prog. Brain Res. **165**, 323–346 (2007)
42. G.E. Loeb, Overcomplete musculature or underspecified tasks? Mot. Control **4**(1), 81–83 (2000)

Part IV
Neuromechanics as a Scientific Tool

Having presented and developed fundamental concepts and techniques to characterize the feasible mechanical capabilities of a system and their corresponding feasible neural commands, it is now possible to explore and discuss several fundamental tenets and theories of motor control.

Crossing this threshold is by no means trivial, or a simple logical progression. It represents the crossing of a clear border from mathematics-based analysis into biology-based scientific inquiry [1–3]. The former is a bottom-up, inductive approach to building models (be they conceptual, analytical or computational) by combining *laws* of mathematics, physics, and mechanics. The behavior of the model, and the intuition it provides, can be carefully compared against past intuition and concrete experimental data before it is accepted as valid.

Modeling biological systems, on the other hand, remains mostly a top-down deductive approach. Testing hypotheses about biological systems by the scientific method proceeds from observed behavior, to the building of mathematical or conceptual models to obtain predictions that are then tested against additional experimental data.

Neuromechanical models are in fact implementations of scientific hypotheses. But deductive top-down approaches makes the emergent behavior of the model difficult to compare against data, intuition, or even other models, because of the lack of all necessary data, the unobservable nature of some relevant variables, and the sheer complexity of biological systems. Thus, the differences that invariably emerge between model predictions and available experimental data can be attributed to a variety of sources, ranging from the choice of organism and behavior to study, the validity of the modeling approach and model elements, and even the model's numerical implementation. And in fact, compatibility of model predictions with experimental data is not proof of the validity of the model.

The nature of this scientific approach explains why, even when applying sound mathematical and engineering principles, our community has intense debates about neuromechanical function. I now explore and discuss several fundamental tenets and theories of motor control from the neuromechanical perspective developed in this book.

References

1. S.P. Ellner, J.M. Guckenheimer, *Dynamic Models in Biology* (Princeton University Press, Princeton, 2011)
2. J. Milton, T. Ohira, *Mathematics as a Laboratory Tool* (Springer, Berlin, 2014)
3. F.J. Valero-Cuevas, H. Hoffmann, M.U. Kurse, J.J. Kutch, E.A. Theodorou, Computational models for neuromuscular function. IEEE Rev. Biomed. Eng. **2**, 110–135 (2009)

Chapter 9
The Nature and Structure of Feasible Sets

Abstract An engineering perspective is inherently incomplete when applied to science. However, as per the words of Galileo Galilei at the beginning of this book, science is also not complete without a mathematical foundation. Our large community applied this mathematics-based perspective for decades to understand motor control. This has resulted in a large, informative, useful, and fruitful body of work. I now comment briefly on how the neuromechanical framework of this book applies to some current tenets, theories, and debates in motor control. In particular, if we agree that the mechanical principles outlined in this book are relevant to the structure of vertebrate limbs, then the nature and structure of the feasible sets they allow are relevant to their neural control. In this chapter I present brief descriptions of how our community has approached understanding the nature and structure of the high-dimensional feasible activation sets.

9.1 Bounding Box Description of Feasible Sets

As described in Part III, Figs. 8.3 and 8.5 are conceptual representations of the family of neural activation patterns (i.e., a convex set containing all valid vectors **a**) that meet all of the constraints that define a particular task. In general, however, this is a subspace embedded in a high-dimensional space. For the moment let us stay, however, with an example where $N = 3$ because it is easy to visualize. In this case we see that the feasible activation set is a line embedded in 3D activation space when satisfying two constraints, as shown in Fig. 8.3. If the limb was meeting three different constraints, then the feasible activation set would be a single point (why?), and there would be no redundancy.

Let us look at this example in more detail using Fig. 9.1. Even for the simplest of cases of a line segment embedded in 3D, the question of how to describe its properties is non-trivial. In Sect. 7.6 we mentioned two potential representations: vertex and linear inequality constraints. In both cases we can list the vertices of the line, otherwise the constraints that define it provide little intuition.

One alternative is to use the *bounding box* to describe the extent and volume of the convex set that defines the feasible activation set. Figure 9.1c, d shows how

© Springer-Verlag London 2016

F.J. Valero-Cuevas, *Fundamentals of Neuromechanics*,
Biosystems & Biorobotics 8, DOI 10.1007/978-1-4471-6747-1_9

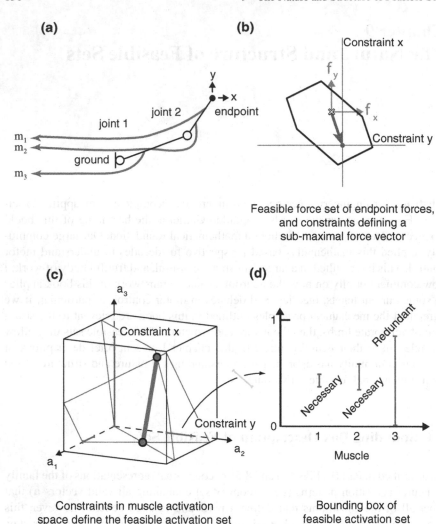

Fig. 9.1 Description of the bounding box of a feasible activation set (adapted from [1]). Panels (**a**) and (**b**) show how the anatomical properties of the limb produce its feasible force set, as described in Chap. 7. If, as described in Chap. 8, a task is defined as producing sub-maximal forces with specific x and y components, then the feasible activation set is the 1D line embedded in 3D shown in (**c**). The *red projections* of the line onto the coordinate axes define the bounding box for the feasible force set as a parallelepiped, whose lengths and ranges are shown as the parallel coordinates in (**d**). Figure adapted with permission from [1]

the bounding box is a linear approximation to the extent of the feasible activation set. This approach has provided intuition about the extent to which muscles can or cannot be used for a particular task. For example, we have used it to describe which muscles are necessary for a particular task, and show that muscle redundancy does not imply robustness to muscle dysfunction. That is, even though there are infinite numbers of solutions in the convex set, they are bounded and not all muscles are equally redundant [1]. Similarly, others have used the bounding box to argue that there remains wide latitude in the choice of muscle coordination strategies when the number of muscles is large [2].

But bounding boxes have clear limitations. For example, creating a feasible activation set with functional constraint planes (like when defining the direction of the output force) means that the dimensionality of the space is reduced by one each time a constraint is added. A cube has dimensionality 3, and the intersection of two planes is a line of dimensionality 1 as shown in Fig. 9.1. The bounding box of a line has a volume equal to the multiplication of the sides of that parallelepiped, but a line has zero volume! In fact, you cannot compare volumes of different dimensions.

In addition, the bounding box shown in parallel coordinates in Fig. 9.1d does not convey those interactions among the different muscles. That is, the orientation of the line imposes strict correlations among muscles, but that information is lost in the bounding box representation.

9.2 Principal Components Analysis Description of Feasible Sets

Principal components analysis, often called PCA, is perhaps the most common means to quantify the structure of subspaces embedded in high dimensions (for an overview see [3, 4]). It is used as a means to estimate the dimensionality of a data set, which is usually collected experimentally. See Fig. 9.2 for an overview. Note that independent components analysis, ICA, is a variant of PCA which does not require finding orthogonal basis vectors but only linearly independent ones [5].

In Fig. 9.2, I show a schematic representation of how PCA is often used as a form of *dimensionality reduction* to test if the data set is can be considered a low-dimensional subspace embedded in the native high-dimensional space.[1]

How do we decide what is the appropriate dimensionality? An extended discussion is provided in [3]. Briefly, as per Fig. 9.2c, the cumulative variance explained by increasing numbers of dimensions (i.e., basis vectors or *principal components*, PCs) is used as a metric of dimensionality. There are as many PCs as there are native dimensions in the data, such as N PCs in $\mathbb{R}^N$, and each PC is associated with a weighting factor σ_i that indicates how much of the variance in the data it explains. When these are normalized to 100%, then cumulatively, $\sum_{i=1}^{N} \sigma_i = 100\%$. In the 3D case, a 2D plane is accepted as a good enough approximation if

---

[1]My preferred use of PCA is as the singular value decomposition of the covariance matrix of the data. In this way the PCs are the left singular vectors in matrix U, and the σ_i are the singular values in the diagonal matrix Σ [3, 4].

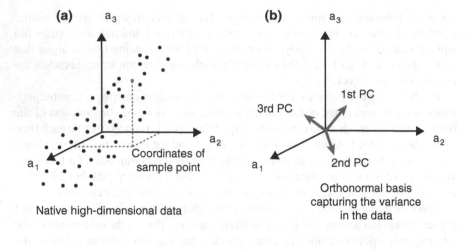

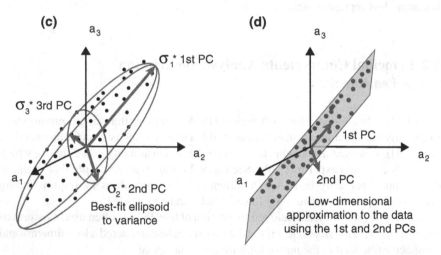

Fig. 9.2 Schematic representation of principal components analysis (PCA) as a means of dimensionality reduction. **a** Given a data set (i.e., a cloud of points), PCA finds the strongest covariances among the points to test if the cloud can be considered a lower-dimensional set (i.e., a plane or a line) embedded in the native high-dimensional space of the points (3D). In this example, PCA would test if the covariance among the points suggests a plane is a good enough approximation. **b** The results of PCA are as many *principal components* (PCs, the basis vectors) as there are dimensions in the native space of the data (3 in this case). These PCs form an orthogonal basis (often orthonormal) with its origin at the centroid of the cloud, and rotated to best capture the variance of the cloud. The elements of the PC basis vectors are sometimes called *loadings*. **c** The strength along each orthonormal PC is given by σ_i, which provides a best-fit ellipsoid as the approximation to the total variance in the data. **d** In PCA the user decides whether or not a few PCs capture the majority of the variance. If so, then the remaining PCs can be left out and the projection of the data onto the lower-dimensional space is accepted as good enough. In this case I decide that σ_1 and σ_2 capture sufficient variance and I project (i.e., flatten) the data onto the plane (*red dots*). That is, I decided that a 2D plane embedded in 3D is an acceptable lower-dimensional approximation to the data and their relevant properties

$$\frac{\sigma_1 + \sigma_2}{\sigma_1 + \sigma_2 + \sigma_3} \geq X\% \ threshold \tag{9.1}$$

where the $X\%$ threshold is a number selected by the user.

In general for N-dimensional data

$$\frac{\sum_{j=1}^{M} \sigma_j}{\sum_{i=1}^{N} \sigma_i} \geq X\% \ threshold$$

$$for \ M < N$$
$$i = 1, \ldots, N$$
$$j = 1, \ldots, M \tag{9.2}$$

In some fields, 60 to 80 % is considered a sufficiently high threshold. When the threshold is met, the interpretation is usually phased as a variant of a statement such as:

Given the N-dimensional data, M principal components suffice to capture at least X % of the variance in the data. Therefore, it is reasonable to say that the data occupy a lower M-dimensional subspace of $\mathbb{R}^N$ [3].

Such dimensionality reduction is nearly always found in experimental behavioral data like electromyograms, joint angles, etc. (see next section). When applied to electromyograms (which are considered physiological estimates of the neural activation command to a muscle [6–8]), this dimensionality reduction is used as an experimental estimate of the dimensionality of the feasible activation set. Note also that the axes of the best-fit ellipsoid in Fig. 9.2c can be considered the dimensions of a properly rotated bounding box as in Fig. 9.1.

The pros and cons of PCA in the context of neural control have been debated extensively. I will visit those debates briefly in the next sections. In the meantime, suffice it to say that PCA is a linear approximation to the dimensionality of experimental data, and its interpretation depends critically on the user's assumptions and goals [3]. In addition, the bases that PCA provides are arbitrary. For example, the mirror image of any result (i.e., where you change the sign of all loadings, thus reverse the sense in which the basis vectors point) is equally valid. This is because you can define the same ellipsoid with a variety of rotated reference frames.

One important limitation of PCA when applied to estimating the dimensionality of neural activation is that the PCs will contain both positive and negative elements. That is, the basis vectors will not exclusively inhabit the positive octant of $\mathbb{R}^N$, where N is the number of independently controllable muscles. That positive octant for three independently controllable muscles is shown in Fig. 7.1. This is contrary to the notion that neural drive is a positive entity.

In response to these limitations, researchers have used *non-negative matrix factorization*, or NMF[2] [9]. This related approach is an iterative optimization process where the goal is to find the lowest-dimensional bases with non-negative basis

---

[2]Sometimes the abbreviation NNMF is also used.

vectors (orthogonal or non-orthogonal) that best explain the variance in the data. In this way, the bases found are compatible with the current notion that neural drive should be a positive entity. Recently, NMF has become the preferred approach to estimate the dimensionality of neural activation data from experiments, or from numerical simulations of experiments [9–13].

Some more subtle issues are that PCA produces results with strictly orthogonal basis vectors[3] and is driven by variance. Both ICA and NMF relax the assumption in PCA of orthogonal axis systems. However, they differ in the following sense: PCA and NMF are variance-based, basis-discovery methods; while ICA seeks to minimize mutual information of sources first and foremost, it examines variance contributions secondarily, and places no constraint on positivity of the basis vectors it finds. As such, ICA is not *doomed-to-succeed*, unlike PCA and NMF, which can always find a least-squares approximation to a cloud of points. Variance and information are related but not identical. The ICA method is more aligned with classical signal processing in an information-based framework, thus it is more sensitive to noise in the signals—which is critical in the case of neural processes.

9.3 Synergy-Based Description of Feasible Sets

The existence and interpretation of *motor, neural* or *muscle synergies*—at times called *motor primitives*—is controversial and has received much attention in the recent literature. Often, synergies are defined as the correlated activation of multiple muscles. Observations of dimensionality reduction in muscle activity and kinematic variables have been interpreted as a strategy the nervous system uses to simplify the neural control of muscles [13–28].

That is, neural synergies are seen as means to simplify and address muscle redundancy, and solve the so-called central problem of motor control (see Chap. 5). These hypotheses around the topic of neural synergies are examples of crossing from mathematics-based analysis into biology-based scientific inquiry. But also note that different groups use the term synergy to mean different things. In this brief presentation, I use "synergies" to mean the correlated activity among muscles.

When exploring these concepts, one must also bear in mind that there are ample and excellent data suggesting that the nervous system may not have absolute, independent control over all muscles (see [29, 30] and references therein)—because of both anatomical and neural constraints. Importantly, the role of perception on motor control, and even independence of muscle actions, is an important one that also requires further understanding [31–33].

Importantly, as pointed out in [15, 19, 23, 34], understanding neural control from a synergy-based approach requires that we distinguish between synergies that are *extracted descriptively from data* versus *synergies that are implemented prescriptively* by a controller or the nervous system. As shown in Fig. 8.3 and others, the

---

[3]Recall that a basis need not have orthogonal basis vectors.

feasible activation set has a well-defined, low-dimensional structure given by the biomechanics of the limb and the mechanical constraints defining the task.

Therefore, it is to be expected that experimental data, such as electromyograms, joint angles, etc., will exhibit correlated activity during the performance of a well-defined task. This correlated activity reflects a lower-dimensional structure embedded in the native higher-dimensional space of the variables as shown in Fig. 9.2. The basis that describes the data is therefore a set of descriptive synergies. Moreover, the better defined the task is, the stronger the structure and lower the dimensionality of the feasible activation space will be as pointed out in Chap. 8 and [35].

One of the interpretations of the existence of such a low-dimensional structure as a function of the constraints of the task is from the perspective of controlled versus uncontrolled trajectories in activation space. The *uncontrolled manifold*[4] concept suggests that traversing along the solution space (i.e., moving within the feasible activation set) remains within the null space of the task and therefore does not require corrective action (i.e., it is a *task-irrelevant* control action). Whereas motion away from the feasible activation set (like leaving the embedded plane in 3D) would need to be controlled to prevent failure of the task and requires a *task-relevant* control action [20, 34]. But there are limits to this strict interpretation of corrective action along these different directions [37]. From the uncontrolled manifold perspective, the basis vectors that correspond to each of these subspaces can be considered a form of task-related variables [34], or synergies [21], that are often assigned functional significance.

One can argue that the uncontrolled manifold perspective is a special case of intermittent or drift-and-act control. These concepts and techniques developed for the control of nonlinear systems where instead of having continuous, online control, control action is used only when needed—given thresholds defining a region of tolerance or null spaces [38–40].

The question then is, how can one infer *prescriptive* synergies (i.e., the existence of synergies of neural origin) from experimental data that naturally exhibit *descriptive* synergies? This is the heart of the debate in this area at the moment. I refer the reader to the references cited in this chapter to explore the many different perspectives held in this field.

It is a subtle, hard to prove [15, 19, 41] distinction whether the low dimensionality of the observed data arises naturally from the nervous system meeting the constraints of the task (by whatever means) versus whether the nervous system is prescriptively implementing a low-dimensional basis to meet the constraints of the task (to simplify the muscle redundancy problem or otherwise). This distinction is not purely semantic.

Muscle synergies can and have been interpreted as both hard-wired evolutionary adaptations, and as learned adaptive strategies [15, 21, 29, 42, 43]. Regardless of their origins, many undeniably functional synergies are known to exist, such as complex limb behavior arising from stimulation of the spinal cord and *central pattern*

---

[4]A *manifold* is a type of subspace of a given dimensionality that is locally Euclidean—but not necessarily linear—that can be embedded in a higher dimensional space. Some examples are lines, planes, spheres, toruses, etc. [36].

generators [26, 42]). But there is also evidence that synergies can be learned, shaped, and adapted (e.g., [13, 17, 44]). These debates continue along several lines of research [19, 22, 27].

The temporal variability of performance (and not only the low-dimensional spatial structure of the data) may hold the key to the inference of the true nature and dimensionality of the prescriptive neural controller that implements the constraints of the task over time. After all, dimensionality reduction methods by themselves provide us only with a spatial average[5] of the neural commands as they meet the constraints of the task. Some examples are the ongoing work along directions that emphasize the distinction between hard-wired versus learned synergies, the evolution of synergies during rehabilitation, variability during performance, cortico-spinal interactions, and the response to perturbations [22, 27, 45–48].

While on the subject of synergies, it is important to mention one of their underappreciated features. Namely, that neural synergies essentially reduce the number of independent DOFs for control. From the (usually large) number of independently controlled muscles, to a smaller number of independently controlled groupings of muscle activations. The presence of synergies, by reducing the number of independent DOFs for control, naturally reduces the size and affects the shape of the feasible activation set (think of it as reducing the number of generators in Fig. 7.2)—and therefore compromises the set of tasks that are feasible [27, 49, 50]. Thus, there is a direct trade off between the real reduction of mechanical capabilities of the limb that synergies produce, the potential benefits to simplified control, and modularity for learning and execution speed.

Moreover, the concept of reduction of control complexity in exchange for reduced capabilities is in reality a continuum that is not well understood. Some argue that this is done by the controller [22, 51], at one end. At the other end, one can design the reduced controllability to match a reduced set of capabilities (i.e., set of feasible tasks) [52], or co-evolve the controller and the body to embed some aspects of control in the physical construction of the system [53–55]. We have argued this provides a glimpse into the nature of brain-body co-evolution of the structure of the human hand [56]. Admittedly, we do not know that much about the evolutionary trajectory vertebrates followed to this point, but either extreme of this continuum is equally unappealing and unlikely. Finding the middle ground should be a focus of future research to recapitulate such game-theoretical processes across evolutionary time, and throughout the lifetime of individual organisms.

Similarly, a line of research related to the material of this book is to treat computational models as experimental subjects and thus see how well current experimental techniques can detect the underlying control strategy purposely implemented in the model by the researcher, or evaluate the tradeoffs of alternative synergies (e.g., [10–12, 19, 23, 57]). That is, one can create a model that has independent muscles, and begin to define different synergies to see their effects.

---

[5]Spatial in the sense that PCA, NMF, etc. provide the low dimensional structure in 'activation space.'

My own perspective is that the question is not *whether* the observed neural commands will inhabit a low-dimensional subspace. Descriptive synergies are to be expected. Rather, the question is *how* the nervous system finds and inhabits that low-dimensional subspace, i.e., implements prescriptive synergies. To make progress I believe the hypotheses about neural synergies should be decoupled from the classical notion of muscle redundancy because, as mentioned in Part II, the control of musculature is likely not as redundant as that notion holds.

In addition, we should focus on understanding the active spatio-temporal processes of how the nervous system finds, inhabits, traverses, and remembers the feasible activation set to meet the requirements of real-world tasks [27, 37, 58–60].

Thus the nature of motor control may be more related to exploration-exploitation strategies driven by trial-and-error with fast and slow learning gradients to characterize and remember the feasible activation set for a task (i.e., a family of good-enough solutions). There are several such strategies that have been producing good results where strict single-line optimization methods are brittle or fail to converge. See Sect. 10.2. For example the work of Lipson, Bongard, and colleagues [53, 61, 62]; and Schaal and colleagues [63–65] has used trial-and-error learning combined with model-free or semi-model free approaches (i.e., they do not require precise models or the full equations of motion of the body being controlled) to evolve complex behavior, or design controllers for complex behavior. It then starts to become biologically feasible that similar memory, interpolation, and extrapolation approaches may be at the root of biological control.

Similarly, a stochastic, Bayesian or Bayesian-like approach of motor control would provide a very plausible view—or metaphor—to how neurons could implement such controllers [66, 67]. This is different from the dominant thinking emphasizing optimization of deterministic systems to find unique solutions. The limits of applying the concepts of optimization to motor control have been discussed elsewhere [68, 69].

My hope is that this book will provide a set of conceptual and computational tools to better understand the dimensionality and structure of feasible activation sets. It is only when we remove the non-neural (i.e., anatomical and task-dependent) contributors to the structure of the these subspaces that we can begin to focus better on the neural contributors to how the nervous system meets those constraints. Along these lines, I present some additional examples of how we can begin to characterize the structure of feasible activations sets in high dimensional spaces in the context of the multiple and compounding constraints of real-world tasks.

9.4 Vectormap Description of Feasible Sets

I will present a brief introduction to a recent approach to visualize the structure of feasible sets. This is described in detail in [70]. Consider the model of the hindlimb

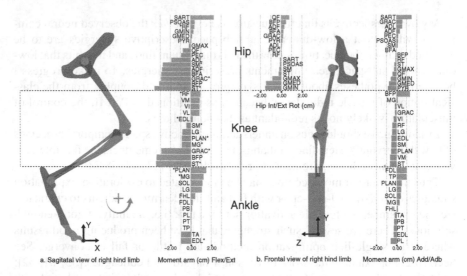

Fig. 9.3 Moment arms of the 7 DOF, 31 muscle cat hindlimb model [2]. Figure adapted with permission from [70]

of a cat (*Felis catus*) with 31 muscles actuating 7 kinematic DOFs from the hip to the ankle,[6] Fig. 9.3.

One application of the techniques described in this book was to find feasible force sets[7] for models of the hindlimb of the cat [71]. A subsequent study explored the bounding box of the feasible activation set for force in one 3D direction [2]. They found that, as the magnitude of the force in that direction increased, the feasible ranges for many muscles remain large even when approaching maximal feasible force.

We studied the bounding box of the feasible activation sets for force production in every 3D direction, also at different force magnitudes [70] because, as motivated by the previous section, understanding neural control strategies requires that we know as much as possible about the global properties and structure of feasible activation sets.

In real life, the neural control of musculature must regulate the magnitude *and the direction* of force vectors, as in locomotion [72] and manipulation [73]. Therefore, it is of interest how the structure of the feasible activation set changes as you vary the direction of force production.

In response to this need, we developed the *vectormap* approach as shown in Fig. 9.4 [70]. This technique allows one to visualize the properties of the enclosed feasible set more intuitively. More importantly, by mapping the properties of the enclosed feasible set onto the surface of the sphere, several calculations can be done

---

[6]The general model has 7 DOFs, but in this analysis hip ad- ab-duction was frozen so the model is really a 6 DOF model.

[7]That is, constraints were added to enforce that the torque elements of the output wrench be zero.

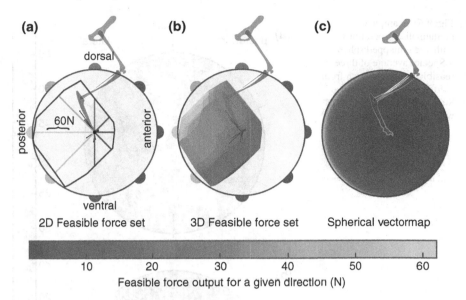

Fig. 9.4 Vectormap visualization for a cat hindlimb in a static force production task. **a** Take the example of a 2D feasible force set, where that polygon is enclosed in a circle. The *thin black lines* emanating from the origin are the lines of action of each of the 31 muscles. The distance from the origin to the boundary of the feasible force set (i.e., the maximal force in that direction) is assigned a color, *blue* for small and *yellow* for large, as shown for the eight rays emanating from the origin. That color-code is vectormapped onto that point on the circle as shown for the posterior, anterior, dorsal and ventral directions, and 4 others in between. **b** The same can be done for a 3D feasible force set that is a polyhedron enclosed in a sphere (only the cross-section of the sphere is shown). **c** The color on the surface of the sphere now contains all the information of the enclosed feasible force set—but in a more intuitive way that can be compared across directions and 3D feasible sets. Adapted with permission from [70]

on that manifold. For example, one can find the difference, sum, or average of two such feasible sets. For example, Fig. 9.5 shows the average and standard deviation maximal feasible force from three different cat models with different limb anatomy and moment arms. The vectormap approach is novel and useful because it allows us to generate cross-species, inter-species, and inter-task comparisons.

This same methodology can be used to represent feasible activations. Take for example the feasible activations associated with the spherical vectormap of the feasible force set shown in Fig. 9.4c. For each 3D direction of the feasible force set, that muscle is associated with a particular activation level. These values are found from the optimal, unique coordination patterns that produce maximal force in each of those directions mapped onto the spherical vectormap of the feasible force set. If you collect all such unique values, they are in fact a 3D object, whose vectormap can be shown in 3D. Consider the case of one single muscle, say the *vastus lateralis*, as shown in the top row of Fig. 9.6. The large vectormap at the far right shows the

Fig. 9.5 Example of mathematical operations with vectormapped spheres. **a** Species average of three feasible force sets, each from a different cat model. **b** Standard deviation. Adapted with permission from [70]

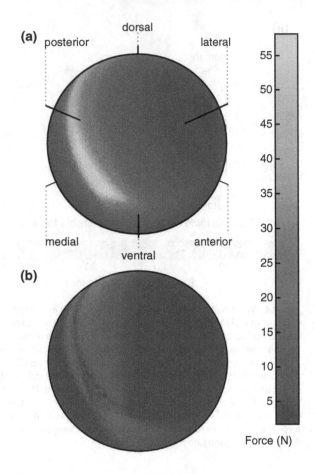

distribution of the unique muscle activation of that muscle for *every* maximal force in 3D.

Now consider the activation of that same muscle for sub-maximal forces. As described in prior chapters and [1, 2], force production in each specific 3D direction has an entire feasible activation space, like the line that produces a single vertex in output space in Fig. 7.6. That feasible activation space is a polygon embedded in N-dimensional space, where N is the number of muscles. As such, it can be described by its bounding box, and that bounding box has a minimal and maximal value for every muscle. Let us take the maximal values for 1 muscle. If we calculate what that maximal value is in every possible 3D direction, we can project that value onto a spherical vectormap. Those vectormaps are shown for 5 different sub-maximal force levels (50, 60, 70, 80, and 90 %) in the top rows associated with each of 3 muscles in Fig. 9.6. We have done the same for the minimal value of the bounding box of the feasible activation set as shown in the bottom rows associated with each of 3 muscles.

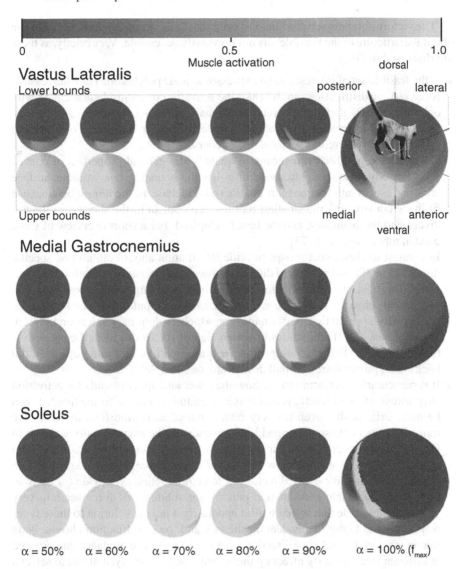

Fig. 9.6 Vectormaps of the feasible activations sets for sub-maximal and maximal force output in every 3D direction. The vectormaps for 5 different sub-maximal force levels (α = 50, 60, 70, 80, and 90 %) for each of 3 muscles are shown, plus the larger vectormap for the maximal feasible force magnitude. The *top* and *bottom rows* are, for each muscle respectively, obtained from the maximal and minimal values of their bounding boxes. Figure adapted with permission from [70]

The vectormap approach provides important advantages, information and intuition about the structure of the feasible sets not previously accessible. Very briefly, as these are discussed in [70]:

- If the feasible set of interest can be expressed as a 3D polyhedron (like the feasible force set), or distributions in 3D (like the activation of a muscle associated with directions in 3D), then the spherical vector map provides a coordinate system (i.e., the surface of the sphere) to combine, operate, or compare feasible sets. The same applies in 2D for circular vectormaps, of course.
- There are many other prior means to quantify the general properties of convex sets. These include straightforward ones like the bounding box, volume, surface area, and aspect ratio. Some more abstract ones include the are largest ball that can fit in the convex set,[8] the smallest ball that can contain it, the shortest or longest distance to the boundary, and the best fit ellipsoid. For a short overview of those used in robotics see [74, 75].
  In contrast to these, vectormaps provide information about both all and specific directions in 3D, and subsets of directions—without smoothing or enforcing symmetry. This has implications to, for example, understanding how phenotypical (i.e., anatomical) changes lead to behavioral changes in output wrench and activation for spatially-specific task performance, on which evolutionary selection may act. Think of how the 3D directions in which highest forces or most efficient activations vary across limbs whose phenotype is adapted to either jump upwards, dig backwards, push forward or pull downward, or example.
- It is particularly interesting to see how the lower and upper bounds for activation vary across all 3D directions and force magnitudes relative to maximal. It can be quite striking that even for very near maximal activation (i.e., at 90%), the range in between these upper and lower bounds can be exceptionally wide, as in Fig. 4. This has been reported in a single direction of force production by Sohn et al. [2], but here we can show the rate of convergence to the unique solution for maximal force for every direction of force production. The wide (or narrow) latitudes in allowable coordination patterns for submaximal force seem to very clearly demonstrate that trying to find and justify a unique solution to these types of problems is highly dependent on the task and the cost function chosen. Note that in other muscles and/or directions this rate of loss of redundancy can proceed at different rates, directly affecting the latitude the nervous system has to select a given coordination strategy.
- Similarly important is the local structure of the feasible activation within and across spherical vectormaps. An example that comes to mind looking at the 3 muscles shown (and more available online), is that the interaction between limb mechanics and task constraints leads to irregular and complex levels of activation across 3D directions of force production for a a given muscle. While there are areas of smooth, almost constant level of activation, there are others in which there are abrupt changes with small changes in direction. Even for sub-maximal forces and

---

[8]Sometimes called the *inscribed ball*. The term *ball* is used in geometry to mean a sphere in arbitrary dimensions. The *circumscribed ball* is the one in which the convex set fits.

in directions important for propulsion. This counters the widespread view that muscles are engaged in a manner consistent with spatially smooth cosine tuning functions [76].

Taken together, the information obtained from the vectormaps begins to shed light on several under-appreciated aspects of the challenges the nervous system faces when controlling tendon-driven limbs. Take the example of changing the direction of an endpoint force vector. The nervous system must not only control the *net level of activation* across all muscles simultaneously, but also their *simultaneous rate of change*.

As discussed elsewhere [37, 58, 59], this becomes a *spatio-temporal* control problem.[9] That is, the time history of neural activations needed to change the direction of a force vector must accommodate, at the very least, the landscape given by the vectormaps, the physiological restrictions on how quickly the neural drive to a muscle can change, and how quickly muscle force can increase and decrease as per the nature of muscle activation-contraction dynamics [77]. Once again, the same arguments presented in Chap. 8 about compounding of task constraints apply, but now in the expanded spatio-temporal domain. Add to this the constraints on tendon excursions and muscle fiber velocities as described in Chap. 6, and it becomes clearer that the neural control of musculature is likely not as redundant as is currently thought. That is, satisfying more and more real-life constraints shows us that the feasible activation space is smaller than previously thought/modeled.

Vectormaps are, therefore, a useful visualization tool that begins to address the need for computational means to characterize and explore the extent to which mechanical considerations determine the neural control of numerous muscles [1, 15, 19, 44].

9.5 Probabilistic Neural Control

Quantifying the theoretical ranges across which each muscle can be activated is useful, but what can one say about which levels of activation are actually more common or likely in practice?

In Sect. 9.1 I point out some limitations of bounding boxes, and in Sect. 9.4 I mention how the development of vectormaps is motivated by the limitations of other means to characterize and visualize the detailed structure of high-dimensional poly-topes embedded in even higher-dimensions. Having said this, vectormaps also share some of the limitations of bounding boxes.

Therefore, there is still need to develop other methods that can give us additional information and intuition about the internal structure of feasible sets. After all, the nervous system is continually exploring, sampling, learning, refining, implementing, and even possibly optimizing within the interior of these high-dimensional structures.

---

[9]Spatial in the sense of meeting the constraints in the space of neural activations, and temporal in the sense of implementing certain temporal dynamics.

We have recently developed a probabilistic method to characterize the internal structure of such high-dimensional convex sets embedded in even higher-dimensional spaces [78]. This method is best described by example. Consider a same representative 3 muscle limb as in Fig. 7.1, that leads to a 3D feasible activation set that is the intersection of a plane with a unit cube in the positive octant of activations as shown in Fig. 8.2.

The question is, what are the probability density functions of muscle activations that satisfy the constraints? Applying the methodology in Fig. 9.1d would show us the bounds on the activation of each muscle as in [1] or [2], but we would not know which level of activity is most common for each muscle. Similarly, if we were to do PCA on data sampled from that feasible activation set, we would know the best-fit cross-correlations among the data points that give us the orientation of the plane as PCs, as in Fig. 9.2d. And NMF would provide us a more physiologically realistic basis for that plane than PCA—but in either case we would not know their bounds, or which level of activity is most common.

We applied techniques developed in the combined fields of computational geometry and stochastic sampling, where the challenge is to develop computationally efficient methods to sample points from high-dimensional convex polytopes. That is, given a high-dimensional convex polytope, how do you sample points that are strictly, but also uniformly distributed, within it?

The methodology we applied is known as *Hit-and-Run*, as developed by Smith [79]. The algorithm is described graphically in Fig. 9.7. The goal is to efficiently sample points that are uniformly distributed within a convex polytope of arbitrary dimensionality. If the Hit-and-Run algorithm can properly sample from within the convex polytope, then the sampled points are representative of all points in the convex set, and the characteristics and probability distributions of those points describe well the details and statistics of the internal structure of the feasible activation set. The details and statistics are an informative characterization of the family of valid neural activation vectors, or muscle coordination patterns—whose importance was described in Sects. 9.3 and 9.4 above.

Consider the feasible activation set for the 3 muscle limb shown in Fig. 7.1, which is satisfying one constraint equation as in Fig. 8.2. The feasible activation set is the 2D convex polygon embedded in 3D, shown by the shaded plane. We see that the results of the Hit-and-Run algorithm are the probability density functions of the points that create the feasible activation set, Fig. 9.7e, f. Were we to use the bounding box approach, all we would know are the extremes of these distributions. We now see that those bounds apply, but that the distribution of neural activation within those bounds is far from uniform. That is, not all values of activation for each muscle are equally likely.

Increasing or decreasing the target magnitude of force output simply translates the plane in a direction orthogonal to it, much as in in Fig. 5.3 when changing the value of a cost function (why?). Figure 9.7f shows that, as the plane moves toward the back corner of the unit cube (in this view) to represent a new target output force magnitude of 2 N, the probability distributions change. When producing maximal feasible force magnitude, the solution is a unique point as the plane P shrinks to

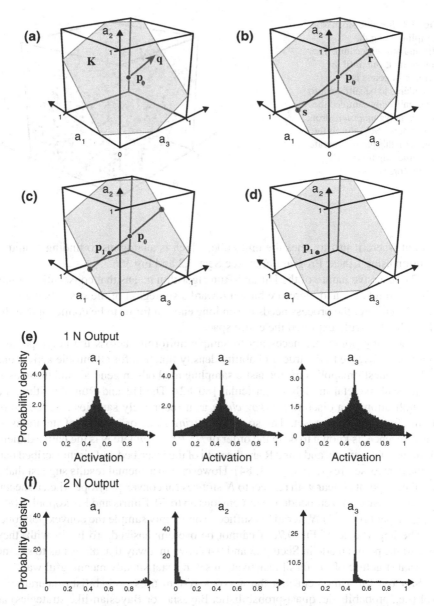

Fig. 9.7 The Hit-and-Run algorithm applied to a convex body K defined as the 2D shaded polygon embedded in 3D. **a** You begin by finding a valid point $p_0 \in K$, and run a linear program in a random vector direction q, (**b**) to find the extreme points r and s at both ends. **c** You then use a uniform probability distribution to sample a point p_1 along the line $\overline{rs}$, and **d** use the point p_1 to repeat step (**a**) iteratively to find the set of points p_i for $i = 0, \ldots, M$, where M is large enough to guarantee convergence to a uniform sampling of K. For an example of such iterative stochastic sampling see Fig. 9.8. **e** The histograms of the sampled points p_i are an approximation to the probability density functions of the convex body K for a 1 N target output force magnitude by the limb in a particular direction. **f** If we change the target magnitude of force output to 2 N, the probability distributions change where now a_2 cannot be fully activated, a_1 moves toward full activation, and the plane's surface area is centrally distributed about a_3

Fig. 9.8 Iterative stochastic sampling using the Hit-and-Run algorithm. The algorithm is run until the number of points is considered to be sufficient to have uniformly sampled the convex polygon, polyhedron, or polytope. Notably, all the points on this plane generate the same magnitude of output force

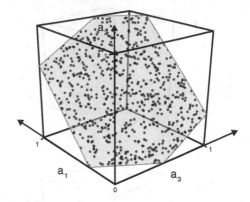

a point where it just touches the unit cube, which is analogous to finding a unique solution using Linear Programming, see Sect. 5.2 and Fig. 9.6.

The recursive nature of the Hit-and-Run algorithm means that consecutive points will be correlated, and therefore biased toward a subregion of the space being sampled. Therefore, the process needs to run long enough for us to be confident that the algorithm has truly explored the entire space.

How many points are necessary to sample uniformly across the polytope, and therefore converge to the true probability density function for all muscle activations a_i? This question applies to stochastic sampling methods in general, and those used for musculoskeletal models in particular [80–82]. The Hit-and-Run algorithm is a generalization of a discrete *Markov chain* as it recursively samples a sequence of points from the convex set. The so-called *mixing time*, or number of iterations to converegence, is known to be in the order of $N^2 R^2 / r^2$, where N is the native dimensionality of the space, and r and R are the radii of the inscribed and circumscribed ball of the convex set, respectively [83, 84]. However experimental results suggest that a number of points linear with respect to N suffices for convex polytopes (i.e., a special class of convex sets) embedded in dimensions up to 40. Emiris and Fisikopoulos [85] suggest that $(10 + \frac{10}{N})\,N$ iterations suffice to uniformly sample the convex polytope.

The importance of Fig. 9.7e, f cannot be overemphasized. To begin with, they extend the points made in Sects. 9.3 and 9.4 above in a way that allows us to see the internal structure of the feasible activation set in a statistically meaningful way.

Suppose for a moment that the nervous system uses probabilistic neural control (i.e., probabilistic, quasi-probabilistic, Bayesian or Bayesian-like strategies) as mentioned in Sect. 9.3. Two enabling—and biologically feasible—elements for that probabilistic approach to work are trial-and-error iterative learning, and memory-based use of probability density functions [37, 66, 67]. Prior to the use of the Hit-and-Run algorithm in this context, such probability density functions could not be described with any detail because of the high-dimensionality of the problem, and the confounds associate from extracting low-dimensional estimates of those spaces using PCA, NMF, etc.

By providing the probability density functions of feasible motor commands for high-dimensional feasible sets, the Hit-and-Run algorithm makes it possible to explore and test the validity of hypotheses emphasizing probabilistic neural control. Several avenues for research come to mind given the state of the art:

- For some muscles, the bounding box may exceptionally misconstrue the internal structure of the feasible activation set—which has important consequences to motor learning. For example, the probability distribution of muscle activations can show strong modes (i.e., narrow and high peaks). Given the high-dimensionality of the problem, the organism can only execute so many trial-and-error iterations during learning, which are likely much too few to completely and exhaustively sample the feasible space of interest. This makes it much more likely that an organism will find those strong modes during learning than *any other region* of the feasible ranges of activation. Thus, the *theoretically* broad ranges of feasible activations for some muscles obtained from the bounding box [1, 2, 70] may have little *practical* bearing on how those tasks are learned and executed.

- If one is to use an exploration-exploitation approach to motor learning and execution, those same strong modes would represent strong local minima or *motor habits*. Habitual control has been proposed based on experimental and empirical data as an alternative to a strict optimization-based approach to motor control [69]. With the Hit-and-Run algorithm now we have the computational means to generate testable hypotheses of how these habits may come about, how they are learned, and how difficult or easy it is to break out of them.

- In fact, falling into such motor habits has been proposed as an intrinsic property of spinal circuitry when learning motor actions in multi-joint, multi-muscle limbs [51]. Thus, the probability density functions produced by the Hit-and-Run algorithm can begin to generate testable hypotheses about motor learning by such simulated populations of neurons.

- Interestingly, those strong probabilistic modes do not necessarily correlate with the activation levels needed for the maximal feasible output (see changes in the modes between 1 and 2 N force output in Fig. 9.7e, f). Therefore, if one is to be efficient (others would say optimal) in the control of their limbs, motor learning needs to proceed from adopting easily-found solutions, to transitioning to less likely subregions of the solutions space. This is of course a proper justification for, and use of, optimization as argued in Sect. 9.3. But that optimization must take place against the strong statistical and memory-based tendency to remain near a strong mode. Only then can the neural system break free of any unavoidable (and useful, I might add) habits formed during the learning process. This may explain why only few humans are able to perform at their optimal levels, and why coaches are so important to elite athletes.

- This line of thinking has consequences to the rehabilitation and restoration of limb function. Neurological conditions of aging, such as stroke, disrupt feasible functions and their associated feasible neural commands, be it by changing the anatomy of the limb, the strength and routing of the tendons, or the independence with which muscles can be controlled. Such disruption, if not destruction, of the

the size and shape of the feasible activation set in older adults will likely require the rebuilding and relearning of existent and or new probability density functions, just when older adults suffer from reduced perceptuo-motor learning rates [86]. Such perspective begins to suggest why rehabilitation in aging adults is so difficult. This suggests particular rehabilitation strategies, and testable hypotheses around them, to exploit knowledge of the statistical structure of feasible function and their associated feasible neural commands.

9.6 Exercises and Computer Code

Exercises and computer code for this chapter in various languages can be found at http://extras.springer.com or found by searching the World Wide Web by title and author.

References

1. J.J. Kutch, F.J. Valero-Cuevas, Muscle redundancy does not imply robustness to muscle dysfunction. J. Biomech. **44**(7), 1264–1270 (2011)
2. M.H. Sohn, J.L. McKay, L.H. Ting, Defining feasible bounds on muscle activation in a redundant biomechanical task: practical implications of redundancy. J. Biomech. **46**(7), 1363–1368 (2013)
3. R.H. Clewley, J.M. Guckenheimer, F.J. Valero-Cuevas, Estimating effective degrees of freedom in motor systems. IEEE Trans. Biomed. Eng. **55**, 430–442 (2008)
4. Wikipedia contributors. Principal components analysis. Wikipedia, The Free Encyclopedia. http://en.wikipedia.org/wiki/Principal_component_analysis. Accessed 29 May 2015
5. A. Hyvärinen, J. Karhunen, E. Oja, *Independent Component Analysis*, vol. 46 (Wiley, New York, 2004)
6. E.R. Kandel, J.H. Schwartz, T.M. Jessell, et al., *Principles of Neural Science*, vol. 4 (McGraw-Hill, New York, 2000)
7. J.V. Basmajian, C.J. De Luca, *Muscles Alive. Muscles Alive: Their Functions Revealed by Electromyography*, vol. 278 (Williams & Wilkins, Baltimore, 1985), p. 126
8. G.E. Loeb, *Electromyography for Experimentalists* (University of Chicago Press, Chicago, 1986)
9. M.C. Tresch, V.C.K. Cheung, A. d'Avella, Matrix factorization algorithms for the identification of muscle synergies: evaluation on simulated and experimental data sets. J. Neurophysiol. **95**(4), 2199–2212 (2006)
10. T.J. Burkholder, K.W. van Antwerp, Practical limits on muscle synergy identification by non-negative matrix factorization in systems with mechanical constraints. Med. Biol. Eng. Comput. **51**(1–2), 187–196 (2013)
11. M.N. Moghadam, K. Aminian, M. Asghari, M. Parnianpour, How well do the muscular synergies extracted via non-negative matrix factorisation explain the variation of torque at shoulder joint? Comput. Methods Biomech. Biomed. Eng. **16**(3), 291–301 (2013)
12. M.K. Steele, M.C. Tresch, E.J. Perreault, Consequences of biomechanically constrained tasks in the design and interpretation of synergy analyses. J. Neurophysiol. **113**(7), 2102–2113 (2015)
13. E. Bizzi, V.C.K. Cheung, The neural origin of muscle synergies. Front. Comput. Neurosci. **7**, 51 (2013). doi:10.3389/fncom.2013.00051
14. L.H. Ting, J.L. McKay, Neuromechanics of muscle synergies for posture and movement. Current Opin. Neurobiol. **17**(6), 622–628 (2007)

15. M.C. Tresch, A. Jarc, The case for and against muscle synergies. Current Opin. Neurobiol. **19**(6), 601–607 (2009)
16. L.H. Ting, J.M. Macpherson, A limited set of muscle synergies for force control during a postural task. J. Neurophysiol. **93**(1), 609–613 (2005)
17. A. d'Avella, E. Bizzi, Shared and specific muscle synergies in natural motor behaviors. Proc. Natl. Acad. Sci. USA **102**(8), 3076–3081 (2005)
18. J.J. Kutch, A.D. Kuo, A.M. Bloch, W.Z. Rymer, Endpoint force fluctuations reveal flexible rather than synergistic patterns of muscle cooperation. J. Neurophysiol. **100**(5), 2455–2471 (2008)
19. J.J. Kutch, F.J. Valero-Cuevas, Challenges and new approaches to proving the existence of muscle synergies of neural origin. PLoS Comput. Biol. **8**(5), e1002434 (2012)
20. M.L. Latash, J.P. Scholz, G. Schoner, Toward a new theory of motor synergies. Motor Control **11**(3), 276 (2007)
21. M.L. Latash, *Synergy* (Oxford University Press, USA, 2008)
22. A. D'Avella, M. Giese, T. Schack, Y.P. Ivanenko, T. Flash, Modularity in motor control: from muscle synergies to cognitive action representation, in *Frontiers in Computational Neuroscience Research Topics* (Frontiers Media SA, 2012)
23. C. Alessandro, I. Delis, F. Nori, S. Panzeri, B. Berret, Muscle synergies in neuroscience and robotics: from input-space to task-space perspectives. Front. Comput. Neurosci. **7**, 43 (2013). doi:10.3389/fncom.2013.00043
24. E. Bizzi, F.A. Mussa-Ivaldi, S. Giszter, Computations underlying the execution of movement: a biological perspective. Science **253**(5017), 287–291 (1991)
25. S.F. Giszter, F.A. Mussa-Ivaldi, E. Bizzi, Convergent force fields organized in the frog's spinal cord. J. Neurosci. **13**(2), 467–491 (1993)
26. S. Giszter, V. Patil, C. Hart, Primitives, premotor drives, and pattern generation: a combined computational and neuroethological perspective. Prog. Brain Res. **165**, 323–346 (2007)
27. S.F. Giszter, Motor primitives–new data and future questions. Current Opin. Neurobiol. **33**, 156–165 (2015)
28. M. Berniker, A. Jarc, E. Bizzi, M.C. Tresch, Simplified and effective motor control based on muscle synergies to exploit musculoskeletal dynamics. Proc. Natl. Acad. Sci. **106**(18), 7601–7606 (2009)
29. M.H. Schieber, M. Santello, Hand function: peripheral and central constraints on performance. J. Appl. Physiol. **96**(6), 2293–2300 (2004)
30. H. van Duinen, S.C. Gandevia, Constraints for control of the human hand. J. Physiol. **589**(23), 5583–5593 (2011)
31. F. Mechsner, D. Kerzel, G. Knoblich, W. Prinz, Perceptual basis of bimanual coordination. Nature **414**(6859), 69–73 (2001)
32. J. Ren, S. Huang, J. Zhang, Q. Zhu, A.D. Wilson, W. Snapp-Childs, et al., The 50s cliff: a decline in perceptuo-motor learning, not a deficit in visual motion perception. PLoS ONE **10**(4), e0121708 (2015). doi:10.1371/journal.pone.0121708
33. N. Kang, J.H. Cauraugh, Bimanual force variability in chronic stroke: With and without visual information. Neurosci. Lett. **587**, 41–45 (2015)
34. J.P. Scholz, G. Schöner, The uncontrolled manifold concept: identifying control variables for a functional task. Exp. Brain Res. **126**, 289–306 (1999)
35. G.E. Loeb, Overcomplete musculature or underspecified tasks? Motor Control **4**(1), 81–83 (2000)
36. M. Spivak, *Calculus on Manifolds*, vol. 1 (WA Benjamin, New York, 1965)
37. K. Rácz, F.J. Valero-Cuevas, Spatio-temporal analysis reveals active control of both task-relevant and task-irrelevant variables. Front. Comput. Neurosci. **7**, 155 (2013). doi:10.3389/fncom.2013.00155
38. J. Milton, T. Insperger, G. Stepan, Human balance control: dead zones, intermittency, and micro-chaos. *Mathematical Approaches to Biological Systems* (Springer, Berlin, 2015), pp. 1–28

39. L.A. Elias, R.N. Watanabe, A.F. Kohn, Spinal mechanisms may provide a combination of intermittent and continuous control of human posture: Predictions from a biologically based neuromusculoskeletal model. PLoS Comput. Biol. **10**(11), e1003944 (2014)
40. J.G. Milton, Intermittent motor control: the "drift-and-act" hypothesis. *Progress in Motor Control* (Springer, Berlin, 2013), pp. 169–193
41. A. de Rugy, G.E. Loeb, T.J. Carroll, Are muscle synergies useful for neural control? Front. Comput. Neurosci. **7**, 19 (2013). doi:10.3389/fncom.2013.00019
42. S. Grillner, Control of locomotion in bipeds, tetrapods, and fish. *Comprehensive Physiology* (John Wiley & Sons, 2011), http://onlinelibrary.wiley.com/doi/10.1002/cphy.cp010226/abstract
43. K. Rácz, D. Brown, F.J. Valero-Cuevas, An involuntary stereotypical grasp tendency pervades voluntary dynamic multifinger manipulation. J. Neurophysiol. **108**(11), 2896–2911 (2012)
44. L.H. Ting, S.A. Chvatal, S.A. Safavynia, J.L. McKay, Review and perspective: neuromechanical considerations for predicting muscle activation patterns for movement. Int. J. Numer. Methods Biomed. Eng. **28**(10), 1003–1014 (2012)
45. F.J. Valero-Cuevas, M. Venkadesan, E. Todorov, Structured variability of muscle activations supports the minimal intervention principle of motor control. J. Neurophysiol. **102**, 59–68 (2009)
46. W.J. Kargo, S.F. Giszter, Rapid correction of aimed movements by summation of force-field primitives. J. Neurosci. **20**(1), 409–426 (2000)
47. W.J. Kargo, S.F. Giszter, Individual premotor drive pulses, not time-varying synergies, are the units of adjustment for limb trajectories constructed in spinal cord. J. Neurosci. **28**(10), 2409–2425 (2008)
48. T. Drew, J. Kalaska, N. Krouchev, Muscle synergies during locomotion in the cat: a model for motor cortex control. J. Physiol. **586**(5), 1239–1245 (2008)
49. F.J. Valero-Cuevas, A mathematical approach to the mechanical capabilities of limbs and fingers. Adv. Exp. Med. Biol. **629**, 619–633 (2009)
50. R. Balasubramanian, Y. Matsuoka, Biological stiffness control strategies for the anatomically correct testbed (act) hand, in *IEEE International Conference on Robotics and Automation. ICRA 2008* (IEEE, 2008), pp. 737–742
51. G. Raphael, G.A Tsianos, G.E Loeb, Spinal-like regulator facilitates control of a two-degree-of-freedom wrist. J. Neurosci. **30**(28), 9431–9444 (2010)
52. D.E. Koditschek, Task encoding: toward a scientific paradigm for robot planning and control. Robot. Auton. Syst. **9**(1), 5–39 (1992)
53. H. Lipson, J.B. Pollack, Automatic design and manufacture of robotic lifeforms. Nature **406**(6799), 974–978 (2000)
54. R. Pfeifer, J. Bongard, *How the Body Shapes the Way We Think: A New View of Intelligence* (MIT Press, Cambridge, 2006)
55. R.A. Brooks, Artifical life and real robots, in *Toward a Practice of Autonomous Systems: Proceedings of the First European Conference on Artificial Life* (1992), p. 3
56. F.J. Valero-Cuevas, J.W. Yi, D. Brown, R.V. McNamara, C. Paul, H. Lipson, The tendon network of the fingers performs anatomical computation at a macroscopic scale. IEEE Trans. Biomed. Eng. **54**, 1161–1166 (2007)
57. L.H. Ting, S.A. Kautz, D.A. Brown, F.E. Zajac, Phase reversal of biomechanical functions and muscle activity in backward pedaling. J. Neurophysiol. **81**(2), 544–551 (1999)
58. J.B. Dingwell, J. John, J.P. Cusumano, Do humans optimally exploit redundancy to control step variability in walking? PLoS Comput. Biol **6**(7), e1000856 (2010)
59. K.G. Keenan, V.J. Santos, M. Venkadesan, F.J. Valero-Cuevas, Maximal voluntary fingertip force production is not limited by movement speed in combined motion and force tasks. J. Neurosci. **29**, 8784–8789 (2009)
60. M. Venkadesan, F.J. Valero-Cuevas, Neural control of motion-to-force transitions with the fingertip. J. Neurosci. **28**, 1366–1373 (2008)
61. J. Bongard, V. Zykov, H. Lipson, Resilient machines through continuous self-modeling. Science **314**, 1118–1121 (2006)

62. J. Rieffel, F.J. Valero-Cuevas, H. Lipson, Morphological communication: exploiting coupled dynamics in a complex mechanical structure to achieve locomotion. J. Royal Soc. Interface (2009) (In Press)
63. E. Theodorou, E. Todorov, F.J. Valero-Cuevas, Neuromuscular stochastic optimal control of a tendon driven index finger model, in *American Control Conference (ACC)* (IEEE, 2011), pp. 348–355
64. E. Theodorou, J. Buchli, S. Schaal, A generalized path integral control approach to reinforcement learning. J. Mach. Learn. Res. **11**, 3137–3181 (2010)
65. M. Kalakrishnan, J. Buchli, P. Pastor, M. Mistry, S. Schaal, Learning, planning, and control for quadruped locomotion over challenging terrain. Int. J. Robot. Res. **30**(2), 236–258 (2011)
66. K.P. Körding, D.M. Wolpert, Bayesian integration in sensorimotor learning. Nature **427**(6971), 244–247 (2004)
67. T.D. Sanger, Distributed control of uncertain systems using superpositions of linear operators. Neural Comput. **23**(8), 1911–1934 (2011)
68. G.E. Loeb, Optimal isn't good enough. Biol. Cybern. **106**(11–12), 757–765 (2012)
69. A. De Rugy, G.E. Loeb, T.J. Carroll, Muscle coordination is habitual rather than optimal. J. Neurosci. **32**(21), 7384–7391 (2012)
70. F.J. Valero-Cuevas, B.A. Cohn, H.F. Yngvason, E.L. Lawrence, Exploring the high-dimensional structure of muscle redundancy via subject-specific and generic musculoskeletal models. J. Biomech. **48**(11), 2887–2896 (2015)
71. J.L. McKay, T.J. Burkholder, L.H. Ting, Biomechanical capabilities influence postural control strategies in the cat hindlimb. J. Biomech. **40**(10), 2254–2260 (2007)
72. F.C. Anderson, M.G. Pandy, Dynamic optimization of human walking. J. Biomech. Eng. **123**(5), 381–390 (2001)
73. F.J. Valero-Cuevas, N. Smaby, M. Venkadesan, M. Peterson, T. Wright, The strength-dexterity test as a measure of dynamic pinch performance. J. Biomech. **36**, 265–270 (2003)
74. J.M. Inouye, J.J. Kutch, F.J. Valero-Cuevas, A novel synthesis of computational approaches enables optimization of grasp quality of tendon-driven hands. IEEE Trans. Robot. **28**(4), 958–966 (2012)
75. A.T. Miller, P.K. Allen, Graspit! a versatile simulator for robotic grasping. IEEE Robot. Autom. Mag. **11**(4), 110–122 (2004)
76. E. Todorov, Cosine tuning minimizes motor errors. Neural Comput. **14**(6), 1233–1260 (2002)
77. F.E. Zajac, Muscle and tendon: properties, models, scaling, and application to biomechanics and motor control. Crit. Rev. Biomed. Eng. **17**(4), 359–411 (1989)
78. F.J. Valero-Cuevas, B.A. Cohn, M. Szedlák, K. Fukuda, B. Gärtner, Structure of the set of feasible neural commands for complex motor tasks, in *37th Annual International Conference of the IEEE Engineering in Medicine and Biology Society*, Milan, Italy, August 2015. (IEEE Engineering in Medicine and Biology Society, 2015)
79. R.L. Smith, Efficient monte carlo procedures for generating points uniformly distributed over bounded regions. Op. Res. **32**(6), 1296–1308 (1984)
80. V.J. Santos, F.J. Valero-Cuevas, A Bayesian approach to biomechanical modeling to optimize over large parameter spaces while considering anatomical variability, in *Conference Proceedings of IEEE Engineering in Medicine & Biology Society*, vol. 6 (2004), pp. 4626–4629
81. V.J. Santos, C.D. Bustamante, F.J. Valero-Cuevas, Improving the fitness of high-dimensional biomechanical models via data-driven stochastic exploration. IEEE Trans. Biomed. Eng. **56**, 552–564 (2009)
82. F.J. Valero-Cuevas, H. Hoffmann, M.U. Kurse, J.J. Kutch, E.A. Theodorou, Computational models for neuromuscular function. IEEE Rev. Biomed. Eng. **2**, 110–135 (2009)
83. M. Dyer, A. Frieze, R. Kannan, A random polynomial-time algorithm for approximating the volume of convex bodies. J. ACM (JACM) **38**(1), 1–17 (1991)
84. L. Lovász, Hit-and-run mixes fast. Math. Program. **86**(3), 443–461 (1999)
85. I.Z. Emiris, V. Fisikopoulos, Efficient random-walk methods for approximating polytope volume. (2013), arXiv preprint arXiv:1312.2873
86. R.O. Coats, A.D. Wilson, W. Snapp-Childs, A.J. Fath, G.P. Bingham, The 50s cliff: perceptuomotor learning rates across the lifespan. PloS ONE **9**(1), e85758 (2014)

Chapter 10
Implications

Abstract This book is deliberately a short introduction to the mathematical and anatomical foundations of neuromechanics. My hope is that you will take these concepts and challenge, modify, extend, and leverage them to advance the science of neuromuscular control and its related areas, such as robotics, musculoskeletal modeling, computational neuroscience, rehabilitation, and evolutionary biology. Having established a common language, conceptual framework, and computational repertoire, I discuss several implications of this neuromechanical perspective. My intent is that my presentation of several issues, research directions, tenets, and debates, however brief, will inspire and encourage you in your research.

10.1 Muscle Redundancy

As we arrive to this last chapter, it is important for me to return to the words of Galileo Galilei at the beginning of this book, and my motivation and goals as stated in Chap. 1.

I began by questioning whether we should continue to think along the lines that muscle for net joint torque production is the 'central problem' of motor control [1]—interpreted as one of neural computation to select specific commands to muscles from the many (infinite, in fact) options allowed by the many muscles in vertebrate limbs. I then argued that such a perspective is paradoxical with respect to the evolutionary process and clinical reality.

Some of us have argued along these lines, stating that muscle abundance is a good thing [2–4], in particular because only when you have multiple muscles can you satisfy the multiple functional constraints of real-world tasks [5, 6]. But this argument resembles a tautology—and evolution does not work on the basis of tautologies. The evidence that we are versatile with many muscles cannot be taken as proof that having many muscles is an evolutionary strategy. The evidence suggests, at best, a strong association between many muscles and evolutionary success (animals never stop evolving, though). But muscles are expensive in the short term and having too many of them can become an unsustainable trait in changing ecological circumstances. Also, muscle excursions are overdetermined. Thus having many muscles exacerbates the

© Springer-Verlag London 2016   159
F.J. Valero-Cuevas, *Fundamentals of Neuromechanics*,
Biosystems & Biorobotics 8, DOI 10.1007/978-1-4471-6747-1_10

risk of disrupted movements due to inaccurate control. Therefore, we need to develop a more fundamental approach (explanation?) rooted in mathematics, as suggested by Galileo. If nothing else, such an approach will allow us to decide how to best learn from organisms to create truly versatile robots.

In my mind, the lack of a formal mathematical approach to the high-dimensionality of the problem of neural control of tendon-driven limbs has forced us to rely on deductive scientific reasoning, (over?) simplified computational models, and conservative robot designs [7]. Biological limbs and nervous systems are so complex that a definitive answer is unlikely to be found in the near future. Without such a *'mathematical language,'* we run the risk, as Galileo put it, of *'wander[ing] in vain through a dark labyrinth.'*

Hence my motivation to write this book: to present the early development (there is still much to be done!) of a computational approach that begins to address the high-dimensionality of the systems of interest, and in so doing allows us to understand the structure of feasible neural commands in the context of the feasible mechanical capabilities of tendon-driven limbs. In the words of Galileo, once again, we should seek to *"Measure what is measurable and make measurable what is not."*

So, what have we learned about the questions posed in Chap. 1? There I asked:

- Why would organisms evolve, encode, grow, maintain, repair, control unnecessarily many muscles when a simpler musculoskeletal system would suffice, and thus have phenotypical and metabolic advantages?
  To this I say, as mentioned above, only when you have multiple muscles can you satisfy the multiple functional constraints of real-world tasks. But having many muscles also requires that we find solutions in high-dimensional activation spaces. To this challenge, the techniques of this book formalize the notion that, while high-dimensional, the feasible activation spaces are very highly structured. That structure enables probabilistic neural control as discussed in Sect. 9.5. This is a distinct alternative to cost function-based optimization approaches that are the dominant perspective today.
- Why would musculoskeletal systems evolve in such a way as to require the nervous system to solve an optimization problem with infinite solutions?
  No one proposes that the nervous system works like a computer. But as explained in this book, optimization methods are wildly successful in many large-scale engineering applications. Therefore, we use them to do our science [7]. I underscore, see for example Sect. 8.1, that it is important that we use—and see—such computational approaches purely as a metaphor. This is hard to do when they are the only widely available techniques, and computational packages like SAGE or MATLAB make them so handy to use. That is why this book is dedicated to proposing an alternative computational geometry approach so that we can begin to explore some, perhaps, more biological plausible approaches to controlling multi-DOF limbs with multiarticular muscles.
- Why do people seek clinical treatment for measurable dysfunction even after injury to a few muscles or mild neuropathology? And, which muscle would you donate to improve or simplify your neural control?

The book makes the strong case that real-life tasks are achieved under numerous and compounded spatio-temporal constraints. Tendon excursions are overdetermined and require accurate temporal control. And as mentioned in Part III, every muscle contributes in unique ways to the dimensionality, size and shape of the feasible activation and the feasible output wrench sets.[1] Therefore, every muscle contributes in unique ways to our ability to meet real-life spatio-temporal constraints.

This mathematical and geometric perspective serves to ground the arguments of muscle abundance, feasibility of task constraints, robustness, and evolutionary success. The computational tools and testable hypothesis that emerge from this perspective help us understand the co-evolution of brain and body in the context of physical versatility.

In fact, these tools can directly illuminate the global interactions among limb geometry, tendon routings, muscle strengths, feasible neural commands, and feasible physical tasks in a way not possible with the current dominant emphasis on muscle redundancy and optimization. These tools and this neuromechanical perspective enable and motivate us to recapitulate the evolutionary process in simulation—being careful not to invoke teleological arguments or anthropomorphize the evolutionary process.

The converse to this argument of versatility is also true. Everyone knows that trauma, paralysis, or pathological synergies cause disability. No surprises there. But to find out which *specific features of disability* are associated with which *specific features of the trauma, paralysis, or pathological synergies* grounds our approach to understanding disability and developing therapeutic strategies.

- Why is it that, despite our best efforts, engineers are still far from creating robots with versatility comparable to that of vertebrates? And, how can such versatile function arise from muscles that have such nonlinear viscoelastic properties, and sensors that are so delayed and noisy?

The material in this book helps mitigate the tautology mentioned above. Just as in the case of versatility and pathology, it is clear that more is better, and less is worse, respectively. But if we are in the position of building biomimetic, neuromorphic, human-compatible, or otherwise versatile robots, we need to ground the arguments of muscle abundance, feasibility of task constraints, robustness, and evolutionary success. Only then can we begin to intelligently and efficiently learn from organisms.

In particular, we can begin to address the question of how organismal evolution ignores, tolerates, or even exploits the properties of the neuro-musculo-skeletal components of the brain-body system. I presented some examples of this throughout the book. For example, in Sect. 4.7 I am led to conclude that the viscoelasticity of musculotendons (i.e., an undesirable property of actuators

---

[1] Very rarely do we find 2 muscles with identical lines of action and functional capabilities. I can think of the *extensor indicis proprius* and the slip of the *extensor digitorum communis* to the index finger as a potential example because they share the same tendon of insertion. But given the lack of strict independence of the muscle fibers of the latter muscle, perhaps the former muscle is needed to enforce functional independence of the index finger.

for engineers) may provide built-in error tolerance to the overdetermined problem of tendon excursion. This may be a critical compliment to, and enabler of, the neural control of smooth movements. What is complex, easy, desirable, difficult, far-fetched or commonsensical to us may not be that to an organism.

My hope is that regardless of whether you are a neuroscientist, roboticist, engineer, physician, evolutionary biologist, physical and occupational therapist, athlete, etc., you have found it fruitful to explore these concepts and apparent paradoxes. While no one approach can hope to resolve these important questions, I hope to have provided you sufficient background in the context of vertebrate limbs to allow you to work toward creating a unified approach to these questions.

This chapter is dedicated to exploring additional issues and implications that could help you progress in your own research endeavors.

10.2 What This Book Did Not Cover

Allow me to remind you of something important I mentioned in Chap. 3. Namely, that this books focuses on static mechanics and simple limb kinematics because it is most is accessible to readers with a variety of scientific and non-scientific backgrounds. This makes it possible for you to apply the concepts presented in this book to many current problems of interest in neuromechanics—even within the context of a 1-semester course for upper class or early graduate students, and working scientists. The main goal is to introduce you to the high-dimensional neuromechanical properties of multi-DOF limbs driven by multiarticular muscles by exploring their feasible function and their associated feasible neural commands.

This has left out, by necessity, several other important areas of research such as computational neuroscience and neuromechanics in the context of dynamical function. However, the foundations provided here in kinematics, mechanics, mathematics, computational geometry, and neural control will serve you well as a necessary foundation from which to expand. Dynamical function and nonlinear control are not as easily accessible without additional formal training in rigid body dynamics, dynamical systems theory, optimization, control theory, and optimal control [8, 9].

But truth be told, many areas of machine learning, control theory and estimation-detection theory have not yet been applied to the problems of dynamical function in neuromechanics, as described in [7] and references therein. Moreover, even if optimal control theory has been applied widely and can be an adequate formalism to create computational models of neuromechanical function, there remain important challenges and limitations that need to be considered and overcome, such as contact transitions, curse of dimensionality, and constraints on states and controls [10, 11].

In the state of the art study of dynamical function, most approaches often espouse over-simplified versions of the classical notion of muscle redundancy, a torque-driven approach, and simplified anatomy [7]—all of which this book argues against.

Todorov and colleagues (e.g., [12]) have begun to address some of those challenges for tendon-driven robots using model predictive control.[2] Model predictive control is being made possible by novel computational modeling platforms and physics engines that can operate in hyper-time (i.e., many times faster than real time), such as MuJoCo http://www.mujoco.org. This approach is making real progress in the control of complex robotic systems. Whether and how, and to what extent, those approaches and solutions apply to the neural strategies used by vertebrates remains to be seen. But this and other lines of research [10, 13, 14] leverage a probabilistic approach to learning, execution, and control.

These approaches to dynamical function in robotics are very much in line with the thrust of this book to find families of feasible spatio-temporal solutions in high-dimensions. They are a growing alternative to the current emphasis on unique solutions that may lack robustness, but are also unlikely to be found in the organism's lifetime given the so-called *curse of dimensionality*[3] [15]. Therefore, you are now well positioned, and in good time, to integrate the material in this book with recent advances in robot control to advance the study of dynamical function in neuromechanics.

Similarly, this book does not present a detailed description of the fundamental physiological mechanisms that make behavior possible. These span a broad spectrum of topics from muscle mechanics and spinal circuitry, to sensorimotor processing and the hierarchical organization of the peripheral and central nervous systems. Many approaches and references to that material are found throughout this book, especially in Sect. 9.3.

10.3 What's in a Name?

Something that this book did cover extensively, albeit implicitly, is the notion that by now you should be very suspicious of muscle names. This is not simply a semantic problem as I believe that muscle names have actually stunted our understanding of neuromechanics for centuries. Do not get me wrong, muscles must have names. But, assigning them names explicitly or implicitly associated with particular functions has clouded our understanding of neuromechanics.

By now, you realize that—in the context of static force production—a muscle is nothing more than a *generator* that contributes to the feasible joint torque and

---

[2]Model predictive control, or MPC, is a computationally intensive approach that solves multiple versions of the problem for a short time horizon into the future at each time step. It then uses cost and value functions to pick from among the multiple, most successful branches to assemble families of acceptable full trajectories to solve the problem.

[3]In areas such as numerical analysis, optimization, sampling, combinatorics, machine learning, data mining, etc., as the dimensionality of the variables increases, the volume of the space increases so fast that available data become sparse. Thus the amount of data needed to obtain statistically sound and reliable results often grows exponentially with the dimensionality, rendering the problem impractical.

feasible endpoint wrench sets. See Figs. 7.2 and 7.5, for example. In the context of instantaneous limb accelerations, Kuo has shown that neural activations can produce feasible acceleration sets [16]. Muscles also undergo particular excursions as dictated by the time history of joint angles 6.7. As such, their actual function in these high-dimensional spaces is so dependent on context, task constraints, and the activation of all other muscles, that a particular muscle's function cannot be captured by a single name. Yet muscles were each given a single name that, unfortunately, has a very specific and limited functional connotation. How did this come about?

In my mind, muscle naming was an unfortunate historical accident from the way in which anatomy was first studied and standardized in medieval Europe. Given what we know about the work of early anatomists [17, 18] and surgeons [19], many muscle names were assigned according to the joint rotations they induced in a limp cadaver.

Take *flexor digitorum profundus* as an example. It is a deep muscle (i.e., lying under other muscles in the forearm, *profundus*) whose tendon acts on the finger (i.e., *digitorum*). If its tendon is pulled to simulate a muscle contraction, the finger flexes (i.e., *flexor*). This is all correct, of course, in that particular functional context. To be clear, that context was the relationship between tendon excursions and joint rotations. The problem is that people unwittingly took that name to apply to other functional contexts, like that of producing fingertip forces. To see this more clearly, let us look at its so-called antagonist: *extensor digitorum communis*, i.e., the common extensor of the fingers. That muscle extends the fingers when pulled in the limb cadaver. But now let us look at their actions when you use the tips of your index finger and thumb to pick up and hold a pencil. Now palpate the top and bottom of your forearm with your other hand as you increase and decrease the pinch force on the pencil while relaxing all other muscles as much as possible. You will certainly feel the flexor muscles under your forearm contract, but you will also feel your extensors!

Why is this? For centuries people have been misled by the names of the muscles to think that flexors should be active when flexing a joint, and by extension producing forces associated with grasp (after all, to grasp you need to flex the fingers)—but the extensors should not be active as they would naturally extend the finger, open the hand, and drop the object grasped, right? There are two confounding issues here, and I will address muscle naming first, before addressing the relationship between the control of motion versus the control of force in Sect. 10.4.

What is happening here is that your finger and thumb muscles are producing a *static force*, not *unconstrained excursions* as in a limp cadaver. In Figs. 7.5, 8.6, and 8.8, you will be reminded that producing a static force in a particular direction requires the vector sum of multiple muscles, as described in detail in this book, and for the particular case of finger neuromechanics in [20–22]. Given this, there exists a feasible activation set to produce a sub-maximal endpoint wrench that consists of a well-directed fingertip force without any endpoint torques. That feasible activation set is agnostic about the names of the muscles involved as it only imposes combinations of muscle vectors that meet the constraints of the task. Some of them happen to be called flexors and extensors because of their role in other functional contexts [6, 23]. An example of that vector addition is shown in Fig. 10.1.

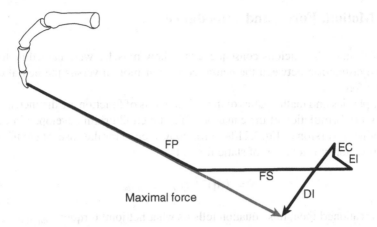

Fig. 10.1 Vector addition needed to produce maximal force with the endpoint of the index finger in the direction of pinching with the thumb. Sub-maximal forces also use interactions among multiple muscles, including the so-called extensor muscles EC and EI [6, 7]. As shown in Fig. 8.1, the names of the muscles used for this task are *flexor digitorum profundus* (FDP), *flexor digitorum superficialis* (FDS), *extensor indicis proprius* (EIP), *extensor digitorum communis* (EDC), and *dorsal interosseous* (DI). Adapted from [24] with permission

A related issue is that muscles have, at times, been categorized as mono-, bi-, or multiarticular depending on whether their tendons cross 1, 2, or more kinematic DOFs, respectively. That distinction is of course clear and important when we discuss issues of tendon excursions, and muscle fiber lengths and velocities [25], such as muscle work, power, etc., as introduced in Chap. 6. However, we must be careful not to generalize.

From the perspective of joint torque production, being a multiarticular muscle simply means that the line of action of that muscle in torque space is not pointing along one of the coordinate axes, as shown in Fig. 7.5 and Chap. 7. But what may be more important functionally is how each muscle contributes to the size and shape of the feasible torque set (or feasible wrench set for that matter, Fig. 7.3). Whether or not the muscle is aligned with the coordinate axes in torque space may or may not matter. In some cases, the line of action of the muscle could be an important evolutionary adaptation for particular tasks that preferentially involve some regions of the torque space [26, 27]. These issues remain relatively unexplored. My hope is that the perspective and techniques presented in this book will enable the study of such co-evolutionary adaptations between brain and body.

10.4 Motion, Force, and Impedance

One of the more pernicious consequences of how muscles were named is that they blur the distinction between the neural control of motion versus the neural control of static force.

The physics and mathematics of these 2 domains of function are distinct, as can be seen by the formulation of the equations of static equilibrium developed in Chap. 3. I reproduce a version of Eq. 3.21 here for convenience for the case where the output wrench vector **w** is a vector of static force, **f**

$$\mathbf{f} = J(\mathbf{q})^{-T} \tau_{for\ static\ force} \qquad (10.1)$$

As mentioned then, this equation tells us what net joint torques $\tau_{for\ static\ force}$ are needed to produce static force of a particular magnitude in a particular direction.

Now consider the *equations of motion* necessary to find the net joint torques needed to produce movement in a given direction, given the initial conditions of limb posture (**q**), angular velocities $\dot{\mathbf{q}}$, and angular accelerations $\ddot{\mathbf{q}}$

$$M(\mathbf{q})\ddot{\mathbf{q}} + C(\mathbf{q}, \dot{\mathbf{q}})\dot{\mathbf{q}} + N(\mathbf{q}) = \tau_{for\ motion} \qquad (10.2)$$

where M represents the inertial properties of the limb, C represents the Coriolis and centrifugal forces on each limb segment, and N is the gravitational term.

Note that if the governing equations are distinct for static force and motion, then it follows that the neural commands and feasible activation sets must also be different [28–32]. Therefore, to say that a muscle is a finger 'extensor' does not mean that it cannot or will not be used to produce grasp forces, as described in Fig. 10.1. In [31] we went further and showed that the nervous system indeed uses two distinct neural control strategies when producing either static force or movement in a same direction with the fingertip.

Thus, in many people's minds, the function of a muscle is interpreted as per its name, which is biased toward a muscle excursion-to-joint rotation interpretation. However, the role of a muscle for behaviors that alternate or combine motion and force production is so dependent on context, task constraints, and the activation of all other muscles, that its function cannot be captured by a single name—or a single notion.

A few of us have highlighted the distinction between the neural control of motion versus force, both theoretically and by showing experimental data that suggest the nervous system indeed treats them differently (e.g., [28, 31]). But the distinction between the control of motion and the control of force, and its implications to motor control, remain relatively unexplored and not well understood. Chapters 5 and 6 highlight this distinction starkly to encourage you to pursue this topic further.

My opinion is that muscle names have lulled us into a false sense of certainty about what muscles are and do; and in this way they have stunted our drive to understand the multifaceted nature of muscle function for real-world tasks.

The related area of *impedance control* has been a subject of much study and debate over the past 3 decades. Muscles are not pure force generators, but exhibit mechanical impedance where their force has a static and dynamic relation to the magnitude and velocity of imposed stretch [33]. Note that it includes both a *stiffness* (length dependent) and a *viscosity* (velocity dependent) term. Thus, one must be careful to define if one speaks of either or both. There are examples in the literature where stiffness is the focus, yet the more general term impedance is used unnecessarily or unjustifiably.

The question is, how does the nervous system modulate or control limb impedance via passive, reflex, or neural mechanisms? The mechanical impedance of the limb determines its reaction forces in response to perturbations from the environment. Thus choosing the mechanical impedance may be one of the ways the nervous system controls its interactions with the environment during dynamical tasks. Early studies showed that the nervous system is capable of varying the total stiffness and viscosity about a joint over a considerable range [34–36].

Numerous experiments and theoretical analyses have studied the biomechanical and neuromuscular capabilities of the nervous system to regulate the impedance of a limb in a variety of contexts, as it is an attractive framework within which to marry the control of motion and force (e.g., [33, 37–57]). While this is often the case, there are of course counterexamples to impedance control [28, 31].

The contents of this book are relevant to the study of impedance control as well. For example, many studies analyze the stiffness the human arm can produce at its endpoint (i.e., the hand) in reaching-like postures in a horizontal plane in front of a seated subject. One set of experimental findings is that, after some training, the nervous system can regulate to varying degrees the orientation and eccentricity of arm stiffness ellipses to perform a task more reliably and efficiently than before training [37, 41, 42, 48]. Another set of experiments concludes that the nervous system cannot arbitrarily regulate endpoint stiffness, and that it is only able to rotate the orientation of the stiffness ellipsoid around 30° [39, 43, 54].

In a recent study we began to reconcile these conflicting results [58] using a 6 muscle planar arm model to show that the presence of synergies drastically decrease the ability of the nervous system to vary the properties of the endpoint stiffness, and can even preclude the ability to minimize energy. Furthermore, the capacity to minimize energy consumption—when available—can be greatly affected by arm posture.

Those results provide further evidence that the benefits and disadvantages of muscle synergies go hand-in-hand with the structure of the feasible activation set afforded by the mechanics of the limb and task constraints. These insights from computational models help design experiments to elucidate the interplay between synergies and the mechanisms of learning, plasticity, versatility and pathology in neuromechanical systems.

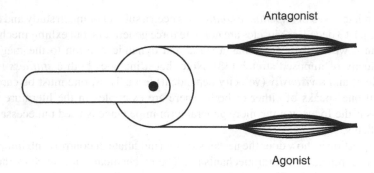

Antagonist

Agonist

Fig. 10.2 An agonist-antagonist pair of muscles acting on a single DOF

10.5 Agonist Versus Antagonist

Another unfortunate historical accident is the didactic habit of using a 1 DOF, 2 muscle limb to teach the integration of neuro-musculo-skeletal elements. Notice I was careful not to do so in the first descriptions of tendon-driven limbs in Figs. 4.7 and 4.8. Such a 'simplest' limb looks as in Fig. 10.2, where 1 muscle is called the *agonist*, and the other *antagonist*. This carries the explicit message and connotation that this agonist-antagonist pair are in direct competition, where the agonist causes an action (i.e., positive torque as per the right hand rule), and the antagonist blocks and inverts that same action. This interpretation is, of course, correct for the 1 DOF, 2 muscle limb. But the problem is that this didactic tool has, much like muscle naming, caused people to have the habit of assigning agonist and antagonist labels and functions to limb muscles—thereby clouding our understanding of muscle function.

In the multi-DOF limbs driven by multiarticular muscles shown in this book, can you find a true agonist-antagonist pair? Take for example the very common concept that the flexor muscles of the fingers are the agonists of a task to produce grasp force. Naturally, the extensor muscles would be the antagonists for all the reasons we discussed about the implications and connotations of muscle names. Now look at Fig. 10.1. Can you really say that about any muscle? Why, or why not?

This concept of agonist is at times carried over to the level of some muscles being the *prime movers* of a task (e.g., [59] and many others). Can you really say that? By now you see that muscles are simply the generators that must be combined to interact with each other and the constraints of the task to define feasible activation and feasible output sets as discussed in Sect. 10.3.

Controlling muscles is much like herding cats in the sense that each wants to go in a different direction, and thus they need to be coordinated as a group. No 1 muscle suffices to meet the functional requirements of real-world tasks. Therefore, the nervous system must make do and coordinate combinations of muscle actions that may or may not be intuitive to us. Once again, controlling muscles in these high-dimensional spaces is so dependent on context, task constraints, and the activation of all other muscles, that its function cannot be captured by a single name, or single connotation.

I would go as far as to say that the terms 'agonist' and 'antagonist' are of no use when working with realistic multi-DOF, multi-muscle limbs like those in Figs. 8.1 and 9.3. They not only fail to capture the functional roles of the muscles, but they also obscure and prevent a useful interpretation of their context-sensitive function.

10.6 Co-contraction

Co-contraction is another term that follows from muscle naming, impedance control, and agonist-antagonist relationships—and is therefore poorly defined and loosely understood. Once again, it emanates from a situation where it can be valid and very precisely defined, just like muscle naming and agonist-antagonists pairs. However, it does not generalize well to multi-DOF limbs driven by multiarticular muscles either.

Consider the agonist-antagonist pair in Fig. 10.2. Here co-contraction has the very clear meaning that, if I stimulate both muscles equally and simultaneously, there will be no net joint torque and thus no motion or force output at the endpoint of the limb. Co-contraction, much like co-activation, is defined for this minimalist limb as the correlated activation of two mutually antagonistic muscles about a joint. So far so good.

But even here we have some semantic problems. The terms muscle co-contraction and joint co-contraction are distinct (depending on whether you care about the state of the muscles, or the state of the joint), but are often used interchangeably. I do not have a particular predilection for either one, but I sidestep this issue by proposing a more formal definition of co-contraction below that applies to joints. Therefore, I will use the term joint co-contraction from now on.

Joint co-contraction is then thought to be one of the principal neural strategies by which the nervous system can regulate the impedance of a joint without changing the output of the limb [33], and therefore a means to modulate the stability of a single DOF (e.g., [60]). This is still correct for the 1 DOF, 2 muscle limb [33].

Where things become less clear is when muscle recordings from so-called *agonist-antagonist pairs of muscles* such as *flexors* and *extensors* are interpreted as evidence that the nervous system is using *co-contraction* of muscles to *stabilize* the multi-DOF, multi-muscle limb.

Let us parse this last sentence carefully. We have established that just because 2 muscles are called flexors or extensors it does not mean that they are an agonist-antagonist pair in the case of multi-DOF limbs driven by multiarticular muscles.

We have also established that increasing limb impedance requires co-contraction [33, 51], but not all co-contraction implies that the system is explicitly, or deliberately, regulating impedance. Consider Fig. 10.3 where we revisit the feasible joint torque set of the familiar limb shown in Fig. 7.5. The generators are the 5 muscles shown in panel Fig. 10.3a. Now suppose you want to produce the net joint torque indicated by the circle in Fig. 10.3b. We have seen that we can achieve this in an infinite number

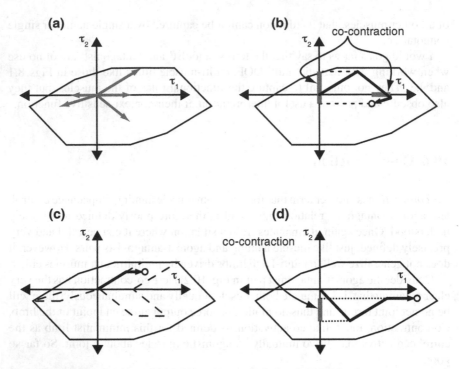

Fig. 10.3 Consider the same multi-DOF, multi-muscle limb as in Fig. 7.5. **a** In torque space, each muscle is a generator that is used to create its feasible joint torque set. **b** Producing a particular net joint torque vector (i.e., *the circle*) requires vector addition of muscle actions in torque space. Such meandering vector additions can, at times, backtrack along the coordinate axes, as shown by the *red lines*. To me this is a rigorous definition of joint co-contraction that generalizes to multi-DOF limbs driven by multiarticular muscles. In this case there is co-contraction at both joints. **c** There are some regions of the feasible torque set—indicated by *dashed lines* for this example—that can be reached without any backtracking, and therefore free of co-contraction. **d** This second example shows co-contraction about the second DOF only. Figure adapted with permission from [61] (color figure online)

of ways (versatility implies redundancy, as in Fig. 8.8), but every solution implies a vector addition of muscle actions in torque space.

In the specific solution shown, that vector addition takes a meandering path but gets the job done (remember, we are herding cats here). What is peculiar about such vector additions is that they at times involve backtracking along the coordinate axes. The red lines on the coordinate axes indicate the amount of joint torque that was produced by some muscles, only to be cancelled by other muscles.

To me this is a more rigorous definition of joint co-contraction that applies to the real-world case of multi-DOF limbs driven by multiarticular muscles, and overcomes the lack of generality of the over-simplistic case of the 1 DOF, 2 muscle limb. This definition provides an objective and quantifiable measure of joint co-contraction. Moreover, it allows us to determine to what extent that joint co-contraction was unavoidable or optional to regulate joint impedance or any other performance metric [51, 58, 62].

Seen from this perspective, joint co-contraction need not be an independently controlled functional goal. It can also be a necessary byproduct of meeting the constraints of the task (i.e., producing the desired net joint torque vector). Note that this distinction is, once again, determined by non-neural factors like tendon routings, the strength and number of muscles, and the constraints of the task—just like synergies and muscle redundancy.

But, you may ask, how does this interpretation hold up if we consider all possible solutions to the problem and not only some specific solutions chosen for convenience? Note that such 'backtracking' will always happen whenever we combine muscle vectors whose lines of actions have a reversal in sign along any of their dimensions. A counter-example is the net joint torque vector that can be achieved without *any* backtracking as shown in Fig. 10.3c. There are, in fact, regions of the feasible joint torque set (highlighted with dashed lines) that contain solutions that can be achieved without any backtracking. In these regions co-contraction is optional as there exist, of course, other solutions that would involve joint co-contraction even though it is avoidable. Thus, the minimal energy solution in those regions of the feasible joint torque set can be achieved without co-contraction at any joint. But outside of these demarcated regions, even the minimal energy solution will involve co-contraction at one of the joints. Therefore a baseline level of co-contraction is no longer optional. Use several vector addition examples to convince yourself of this.

This kind of analysis needs to be applied before attributing the presence, or magnitude, of joint co-contraction to a neural control strategy for something other than producing the required net joint torque vector.

My main points about joint co-contraction and its interpretation therefore are:

- A definition of co-contraction exists that is rigorous and generalizes well to multi-DOF limbs driven by multiarticular muscles.
- The presence or absence of co-contraction may or may not be a optional for the nervous system.
- The options available to the nervous system to control a limb are heavily influenced by the anatomical and physiological properties of the musculoskeletal system.
- Reaching conclusions about co-contraction and its functional purpose in realistic limbs cannot be inferred simply from activity in muscles whose names suggest they are antagonists. A detailed computational model is necessary to arrive at such conclusions, as we and others have have argued in the case of muscle redundancy [2] and synergies [63, 64].

Future studies of the role of co-contraction in the control of neuromechanical function should emphasize the analysis of multi-DOF limbs driven by multiarticular muscles.

10.7 Exercises and Computer Code

Exercises and computer code for this section in various languages can be found at http://extras.springer.com or found by searching the World Wide Web by title and author.

References

1. N.A. Bernstein, *The Co-ordination and Regulation of Movements* (Pergamon Press, New York, 1967)
2. J.J. Kutch, F.J. Valero-Cuevas, Muscle redundancy does not imply robustness to muscle dysfunction. J. Biomech. **44**(7), 1264–1270 (2011)
3. M.L. Latash, The bliss (not the problem) of motor abundance (not redundancy). Exp. Brain Res. **217**(1), 1–5 (2012)
4. G.E. Loeb, Overcomplete musculature or underspecified tasks? Mot. Control **4**(1), 81–83 (2000)
5. K.G. Keenan, V.J. Santos, M. Venkadesan, F.J. Valero-Cuevas, Maximal voluntary fingertip force production is not limited by movement speed in combined motion and force tasks. J. Neurosci. **29**, 8784–8789 (2009)
6. F.J. Valero-Cuevas, F.E. Zajac, C.G. Burgar, Large index-fingertip forces are produced by subject-independent patterns of muscle excitation. J. Biomech. **31**, 693–703 (1998)
7. F.J. Valero-Cuevas, H. Hoffmann, M.U. Kurse, J.J. Kutch, E.A. Theodorou, Computational models for neuromuscular function. IEEE Rev. Biomed. Eng. **2**, 110–135 (2009)
8. R. Shadmehr, S. Mussa-Ivaldi, *Biological Learning and Control: How the Brain Builds Representations, Predicts Events, and Makes Decisions* (MIT Press, Cambridge, 2012)
9. E. Todorov, M.I. Jordan, Optimal feedback control as a theory of motor coordination. Nat. Neurosci. **5**(11), 1226–1235 (2002)
10. E. Theodorou, E. Todorov, F.J. Valero-Cuevas, Neuromuscular stochastic optimal control of a tendon driven index finger model, in *2011 American Control Conference (ACC)* (IEEE, 2011), pp. 348–355
11. E. Theodorou, F.J. Valero-Cuevas, Optimality in neuromuscular systems, in *2010 IEEE Annual International Conference of the Engineering in Medicine and Biology Society (EMBC)* (IEEE, 2010), pp. 4510–4516
12. V. Kumar, Y. Tassa, T. Erez, E. Todorov, Real-time behaviour synthesis for dynamic hand-manipulation, in *2014 IEEE International Conference on Robotics and Automation (ICRA)*, (IEEE, 2014), pp. 6808–6815
13. M. Kalakrishnan, J. Buchli, P. Pastor, M. Mistry, S. Schaal, Learning, planning, and control for quadruped locomotion over challenging terrain. Int. J. Robot. Res. **30**(2), 236–258 (2011)
14. E. Theodorou, J. Buchli, S. Schaal, A generalized path integral control approach to reinforcement learning. J. Mach. Learn. Res. **11**, 3137–3181 (2010)
15. Wikipedia contributors. Basis vectors. Wikipedia, The Free Encyclopedia. http://en.wikipedia.org/wiki/Curse_of_dimensionality. Accessed 7 June 2015
16. A.D. Kuo, F.E. Zajac, Human standing posture: multi-joint movement strategies based on biomechanical constraints. Prog. Brain Res. **97**, 349–358 (1993)
17. F.J. Cole et al. *A History of Comparative Anatomy from Aristotle to the Eighteenth Century* (Macmillan Publisher, London, 1944)
18. A. Vesalius, *De Humani Corporis Fabrica Libri Septem* (Ex officina I. Oporini, Basileae, 1543)
19. R. Van Rijn, The Anatomy Lesson of Dr. Nicolaes Tulp (1632)
20. F.J. Valero-Cuevas, C.F. Small, Load dependence in carpal kinematics during wrist flexion in vivo. Clin. Biomech. **12**, 154–159 (1997)

21. H. van Duinen, S.C. Gandevia, Constraints for control of the human hand. J. Physiol. **589**(23), 5583–5593 (2011)
22. C.E. Wall, A model of temporomandibular joint function in anthropoid primates basedon condylar movements during mastication. Am. J. Phys. Anthropol. **109**(1), 67–88 (1999)
23. F.J. Valero-Cuevas, Predictive modulation of muscle coordination pattern magnitude scales fingertip force magnitude over the voluntary range. J. Neurophysiol. **83**(3), 1469–1479 (2000)
24. F.J. Valero-Cuevas, J.D. Towles, V.R. Hentz, Quantification of fingertip force reduction in the forefinger following simulated paralysis of extensor and intrinsic muscles, J. Biomech. **33**, 1601–1609 (2000)
25. L. Gregoire, H.E. Veeger, P.A. Huijing, S.G.J. van Ingen, Role of mono-and biarticular muscles in explosive movements. Int. J. Sport. Med. **5**(6):301–305, (1984)
26. J.M. Inouye, F.J. Valero-Cuevas, Anthropomorphic tendon-driven robotic hands can exceed human grasping capabilities following optimization. Int. J. Robot. Res. (2013)
27. F.J. Valero-Cuevas, J.W. Yi, D. Brown, R.V. McNamara, C. Paul, H. Lipson, The tendon network of the fingers performs anatomical computation at a macroscopic scale. IEEE Trans. Biomed. Eng. **54**, 1161–1166 (2007)
28. V.S. Chib, M.A Krutky, K.M. Lynch, F.A. Mussa-Ivaldi, The separate neural control of hand movements and contact forces. J. Neurosci. **29**(12), 3939–3947 (2009)
29. R.M. Murray, Z. Li, S.S. Sastry, *A Mathematical Introduction to Robotic Manipulation* (CRC Press, Florida, 1994)
30. V. Squeri, L. Masia, M. Casadio, P. Morasso, E. Vergaro, Force-field compensation in a manual tracking task. PLoS One **5**(6), e11189 (2010)
31. M. Venkadesan, F.J. Valero-Cuevas, Neural control of motion-to-force transitions with the fingertip. J. Neurosci. **28**, 1366–1373 (2008)
32. T. Yoshikawa, *Foundations of Robotics: Analysis and Control* (MIT Press, Cambridge, 1990)
33. N. Hogan, Adaptive control of mechanical impedance by coactivation of antagonist muscles. IEEE Trans. Autom. Control **29**(8), 681–690 (1984)
34. E.R. Kearney, I.W. Hunter, System identification of human joint dynamics. Crit. Rev. Biomed. Eng. **18**(1), 55–87 (1989)
35. J.M. Lanman, Movement and the mechanical properties of the intact human elbow joint. Ph.D. thesis, Massachusetts Institute of Technology (1980)
36. G.I. Zahalak, S.J. Heyman, A quantitative evaluation of the frequency-response characteristics of active human skeletal muscle in vivo. J. Biomech. Eng. **101**(1), 28–37 (1979)
37. E. Burdet, R. Osu, D.W. Franklin, T.E. Milner, M. Kawato, The central nervous system stabilizes unstable dynamics by learning optimal impedance. Nature **414**(6862), 446–449 (2001)
38. E. Burdet, R. Osu, D.W. Franklin, T. Yoshioka, T.E. Milner, M. Kawato, A method for measuring endpoint stiffness during multi-joint arm movements. J. Biomech. **33**(12), 1705–1709 (2000)
39. M. Darainy, N. Malfait, P.L. Gribble, F. Towhidkhah, D.J. Ostry, Learning to control arm stiffness under static conditions. J. Neurophysiol. **92**(6), 3344 (2004)
40. T. Flash, F. Mussa-Ivaldi, Human arm stiffness characteristics during the maintenance of posture. Exp. Brain Res. **82**(2), 315–326 (1990)
41. D.W. Franklin, G. Liaw, T.E. Milner, R. Osu, E. Burdet, M. Kawato, Endpoint stiffness of the arm is directionally tuned to instability in the environment. J. Neurosci. **27**(29), 7705–7716 (2007)
42. D.W. Franklin, U. So, M. Kawato, T.E. Milner, Impedance control balances stability with metabolically costly muscle activation. J. Neurophysiol. **92**(5), 3097 (2004)
43. H. Gomi, R. Osu, Task-dependent viscoelasticity of human multijoint arm and its spatial characteristics for interaction with environments. J. Neurosci. **18**(21), 8965–8978 (1998)
44. N. Hogan, Impedance control: an approach to manipulation, in *American Control Conference* (IEEE, 1984), pp. 304–313
45. N. Hogan, The mechanics of multi-joint posture and movement control. Biol. Cybern. **52**(5), 315–331 (1985)
46. X. Hu, W.M. Murray, E.J. Perreault, Muscle short-range stiffness can be used to estimate the endpoint stiffness of the human arm. J. Neurophysiol. **105**(4), 1633–1641 (2011)

47. H.U. Xiao, W.M. Murray, E.J. Perreault, Biomechanical constraints on the feedforward regulation of endpoint stiffness. J. Neurophysiol. **108**(8), 2083–2091 (2012)
48. A. Kadiallah, G. Liaw, M. Kawato, D.W. Franklin, E. Burdet. Impedance control is selectively tuned to multiple directions of movement. J. Neurophysiol
49. J. McIntyre, F.A. Mussa-Ivaldi, E. Bizzi, The control of stable postures in the multijoint arm. Exp. Brain Res. **110**(2), 248–264 (1996)
50. T.E. Milner, Contribution of geometry and joint stiffness to mechanical stability of the human arm. Exp. Brain Res. **143**(4), 515–519 (2002)
51. F.A. Mussa-Ivaldi, N. Hogan, E. Bizzi, Neural, mechanical, and geometric factors subserving arm posture in humans. J. Neurosci. **5**(10), 2732 (1985)
52. R. Osu, H. Gomi, Multijoint muscle regulation mechanisms examined by measured human arm stiffness and EMG signals. J. Neurophysiol. **81**(4), 1458 (1999)
53. E.J. Perreault, R.F. Kirsch, P.E. Crago, Effects of voluntary force generation on the elastic components of endpoint stiffness. Exp. Brain Res. (Experimentelle Hirnforschung Experimentation cerebrale) **141**(3), 312, (2001)
54. E.J. Perreault, R.F. Kirsch, P.E. Crago, Voluntary control of static endpoint stiffness during force regulation tasks. J. Neurophysiol. **87**(6), 2808 (2002)
55. D. Shin, J. Kim, Y. Koike, A myokinetic arm model for estimating joint torque and stiffness from EMG signals during maintained posture. J. Neurophysiol. **101**(1), 387–401 (2009)
56. S. Stroeve, Impedance characteristics of a neuromusculoskeletal model of the human arm i. Posture control. Biol. Cybern. **81**(5), 475–494 (1999)
57. K.P. Tee, D.W. Franklin, M. Kawato, T.E. Milner, E. Burdet, Concurrent adaptation of force and impedance in the redundant muscle system. Biol. Cybern. **102**(1), 31–44 (2010)
58. J.M. Inouye, F.J. Valero-Cuevas, A novel computational approach helps explain and reconcile conflicting experimental findings on the neural control of arm endpoint stiffness, in *2012 22nd Annual Society for the Neural Control of Movement Conference* (Venice, Italy, 2012)
59. C. Tomberg, M.D. Caramia, Prime mover muscle in finger lift or finger flexion reaction times: identification with transcranial magnetic stimulation. Electroencephalogr. Clin. Neurophysiol. Evoked Potentials Sect. **81**(4), 319–322 (1991)
60. T.E. Milner, Adaptation to destabilizing dynamics by means of muscle cocontraction. Exp. Brain Res. **143**(4), 406–416 (2002)
61. F.J. Valero-Cuevas, An integrative approach to the biomechanical function and neuromuscular control of the fingers. J. Biomech. **38**, 673–684 (2005)
62. R. Balasubramanian, Y. Matsuoka, Biological stiffness control strategies for the anatomically correct testbed *(act)* hand, in *2008 IEEE International Conference on Robotics and Automation (ICRA)* (IEEE, 2008), pp. 737–742
63. J.J. Kutch, F.J. Valero-Cuevas, Challenges and new approaches to proving the existence of muscle synergies of neural origin. PLoS Comput. Biol. **8**(5), e1002434 (2012)
64. M.K. Steele, M.C. Tresch, E.J. Perreault, Consequences of biomechanically constrained tasks in the design and interpretation of synergy analyses. J. Neurophysiol. **113**(7), 2102–2113 (2015)

Appendix A
Primer on Linear Algebra
and the Kinematics of Rigid Bodies

This brief primer provides basic concepts and reference material related to the critical concepts of linear algebra that are needed to work with rigid body kinematics. These concepts are critical to the intuition developed throughout the book. Readers with recent backgrounds in linear algebra may not need this, but others who have not seen linear algebra for some time would benefit from this review. I cannot overemphasize the benefits that come from having working knowledge of these fundamental concepts that are, in fact, within reach for anyone with even cursory exposure to high school algebra. There are multiple texts that cover this material in detail, but I would recommend [1, 2] as first texts, followed by [3] for those interested in a more detailed treatment of the subject in the context of robotics.

A.1 What is a Vector?

There are multiple interpretations of the concept of a *vector* [1]. In this geometric context, I use a vector to mean a *point* in a given *Cartesian space*, Fig. A.1. Note that the invention of the Cartesian coordinate system in the 17th century by René Descartes (1596–1650) (Latinized name: Cartesius) provides a key systematic link between Euclidean geometry and algebra. Wikipedia has an excellent introduction to this topic [4].

In its simplest form, a *vector* is a column of numbers (also called *scalars*) laid out as a 1D *array*) as shown in Eq. A.1 for the case of three elements.

$$\mathbf{v} = \begin{pmatrix} v_1 \\ v_2 \\ v_3 \end{pmatrix} = (v_1, v_2, v_3)^T \qquad (A.1)$$

This equation merits some explanation. First, notice that the vector $\mathbf{v}$ is written in lower case boldface in this text (others use an arrow above the letter, or underline it). The vector in this case is a *column vector*, and written as such. So when written

© Springer-Verlag London 2016
F.J. Valero-Cuevas, *Fundamentals of Neuromechanics*,
Biosystems & Biorobotics 8, DOI 10.1007/978-1-4471-6747-1

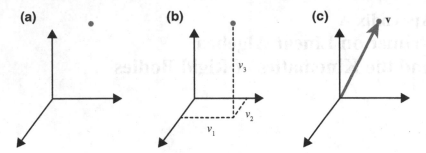

Fig. A.1 Geometric interpretation of a vector. **a** A space can be described using a Cartesian frame of reference. **b** A point in space can then be described by its coordinates in the Cartesian space. **c** A vector is both the location of that point in space, and a description of the magnitude, direction and sense of the displacement from the origin to that point in space

horizontally, it is accompanied by the *transpose* (T) superscript. The transpose of any array is when you make all columns into rows and vice versa. So the transpose of a row is a column, as shown in Eq. A.1. Some vectors are *row vectors*, and thus laid out horizontally, and their transpose is a column vector. The elements (or entries) of the vector **v** are the scalar numbers v_1, v_2, and v_3—written in lower case font to distinguish them from vectors or matrices (see below). I will sometimes use simpler notation to refer to these scalars such as a, b, c, etc. In this text these scalars are most often real numbers that capture distances, forces, activation levels, etc. Last, the dimensionality of the vector is the number of elements. Thus in the case of Eq. A.1, the dimensionality is 3 and we can say that

$$\mathbf{v} \in \mathbb{R}^3 \tag{A.2}$$

which means that the vector **v** belongs to (is in, or $\in$) the set of all arrays that contain three real numbers, $\mathbb{R}^3$.

If we use the example in Fig. A.1—which is in 3D in this case to make it easy to visualize—the entries are the projections of the point along each of the coordinates in 3D Cartesian space that fully specify the location of that point.

But in general, vectors can be of any dimension, call it the number N. This makes their general definition be

$$\mathbf{v} = \begin{pmatrix} v_1 \\ v_2 \\ \vdots \\ v_N \end{pmatrix} = (v_1, v_2, \ldots, v_N)^T \tag{A.3}$$

where

$$\mathbf{v} \in \mathbb{R}^N \tag{A.4}$$

A vector **v**, as you can surmise from Fig. A.1, can also be a physical entity that has a magnitude, direction, and sense. The vector magnitude (i.e., its length) is computed using the Pythagorean theorem, also called the Euclidean norm. In 3D, it is the familiar formula

$$||\mathbf{v}|| = \sqrt{v_1^2 + v_2^2 + v_3^2} \tag{A.5}$$

or more generally in $\mathbb{R}^N$

$$||\mathbf{v}|| = \sqrt{v_1^2 + v_2^2 + \cdots + v_N^2} \tag{A.6}$$

The *vector direction* is, by convention, defined by a *unit vector*: a vector in the same direction as the vector **v**, but that has a magnitude equal to 1. A 'hat' is sometimes used to indicate a unit vector $\hat{\mathbf{v}}$. This normalization is done by dividing the vector, element by element, by its magnitude,

$$\hat{\mathbf{v}} = \frac{\mathbf{v}}{||\mathbf{v}||} \tag{A.7}$$

Therefore, you can think of a vector as a set of instructions that say: start at the origin, and move away from the origin in the direction $\hat{\mathbf{v}}$ a distance $||\mathbf{v}||$. This meaning is complementary to the vector **v** being a location in that Cartesian space. That is, that location **v** is defined by how you get there.

But a vector can take on different meanings depending on the nature of the variables you use as its entries. For example, if the entries of the vector are the individual forces from each of 3 muscles,

$$\mathbf{f} = \begin{pmatrix} f_1 \\ f_2 \\ f_3 \end{pmatrix} \tag{A.8}$$

then the vector **f** is a *muscle force vector* that indicates how the nervous system is coordinating muscle forces [5], as discussed in Sect. 4.4 and throughout the this book.

The beauty of the link between Euclidean geometry and linear algebra (one of the greatest achievements of mathematics) is that these concepts, and the intuition they provide, extend to any number of dimensions. Hence we can, for example, talk about a vector $\mathbf{f} \in \mathbb{R}^{31}$ that describes the coordination of forces across 31 muscles in the hindlimb of a cat [6]. Moreover, as it becomes clear in several parts of this book, $\mathbb{R}^N$ is not simply a set of arrays with N entries. Rather, you can think of it as an N-dimensional *vector space*—be it a physical space, or the 'space' of muscle forces, joint torques, or endpoint forces of a limb as shown in Fig. 7.3.

A *matrix* is also an array of numbers, but with multiple rows and columns. You can think of it as a collection of row or column vectors. Or you can think of a vector as a matrix with only one row or column. When I speak of vectors in this book I often mean column vectors, but discussing the elements of matrices require us to specify whether we are talking about its row or column vectors. I typeset matrices as capital italicized letters.

The dimensionality of a matrix A is given as $A \in \mathbb{R}^{M \times N}$ for a matrix with M rows and N columns. The entries of the matrix are the scalars $a_{i,j}$, for $i = 1, \ldots, M$ and $j = 1, \ldots, N$.

Summarizing details of *vector notation*, *scalar notation* and *indices* for completeness: I use lowercase boldface font for vectors (**v**), and italics for scalars (a). The elements of a vector are scalars with a subscript (v_1). Subscript indices are lowercase italicized letters i, j, etc. to signify a number within a range. Capital letters like M or N to indicate the extremes of the ranges (a_i, for $i = 1, \ldots, N$). A matrix is typeset as a capital, italicized, letter (A). The dimensions of a vector or matrix are given using capital letters like $\mathbf{v} \in \mathbb{R}^N$ or $A \in \mathbb{R}^{M \times N}$, respectively. The letter M need not stand invariably for muscles, nor i for row. They are simply letters to indicate dimensions and indices, and their meaning changes with the context of the material. This unfortunate reuse of variables is not uncommon in mathematical descriptions, as it is assumed that context dictates meaning.

A.2 Frames of Reference

Staying with the geometric interpretation of a vector and vectors spaces, you are probably realizing that the description of the vector **v** depends entirely on the frame of reference used—even though a location in space exists independently of any frame of reference. Thus the location (i.e., its origin o) and orientation of the frame of reference needs to be well defined. Figure A.2 shows one such example where a same vector **v** 'looks' different from each of two frames of reference that have the same origin at the start of the vector. That is, the entries of the vector differ even though the vector starts and ends in the same locations in space.

A *basis* is a set of linearly independent vectors that span a space [7]. But it also contains the fewest basis vectors that span the space, such that $\mathbb{R}^N$ can have bases with at most N basis vectors. Thus the number of basis vectors is the dimension of the space (one for a line, two for a plane, three for a space, etc., see Sect. 9.2). Each basis is not unique as there are multiple equivalent bases that span a given space, thus the user needs to define the basis of choice.

A vector is said to be *linearly independent* [8] of others if it cannot be written as their linear combination (i.e., weighted sum). A classical example is

$$\begin{pmatrix} 1 \\ 0 \\ 0 \end{pmatrix} \neq a \begin{pmatrix} 0 \\ 1 \\ 0 \end{pmatrix} + b \begin{pmatrix} 0 \\ 0 \\ 1 \end{pmatrix} \tag{A.9}$$

as there are no possible a and b scalars to make the equality hold.

In this example, any point in 3D space can be described using linear combinations of the linearly independent basis vectors in Eq. A.9. Thus, those three vectors can be

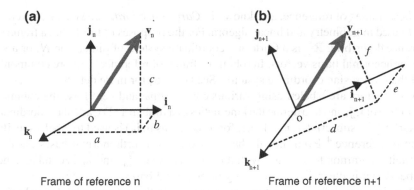

Fig. A.2 A same vector **v** in 3D described using two different frames of reference, labeled **a** n, and **b** n+1. Therefore the vectors are labeled v_n and v_{n+1}, respectively. Both frames of reference have the same origin, o at the start of the vector, but each has a different orientation. They are composed of as many orthonormal (i.e., perpendicular and unit length) basis vectors as there are dimensions of the space; three in this case. Each basis vector is a coordinate axis labeled **i**, **j**, and **k** fixed to the origin with a subscript identifying the frame of reference to which it belongs. Note that the projections onto their respective frames of reference, and therefore the entries of the vectors, differ as shown in Eqs. A.13 and A.14

a basis that is a natural form of a *frame of reference* for that space. But note that the vectors

$$\begin{pmatrix} -6 \\ 0 \\ 0 \end{pmatrix}, \begin{pmatrix} 0 \\ 4 \\ 0 \end{pmatrix}, \begin{pmatrix} 0 \\ 0 \\ 5.2 \end{pmatrix} \tag{A.10}$$

are an equally valid choice of an orthogonal basis for 3D space.

Orthonormal bases are preferred in kinematics, dynamics, and other fields. They are defined by containing basis vectors that are both orthogonal[1] to each other and of unit magnitude, as in Eq. A.9. That is why those vectors look so familiar to you, but those in Eq. A.10 seem strange. Such orthonormal basis vectors are often used as the *orthogonal coordinate axes* and *dimensions* of spaces. Figure A.2 shows two such frames of reference.

Note that the direction and order of the basis vectors of a frame of reference are defined as per the Ampères *right-hand rule*[2]: point your right index finger along the first basis vector, i_n, and rotate your wrist to align it with the second one, j_n. This leaves your right thumb pointing along the third basis vector, k_n.

---

[1] 'Orthogonal' is the formal term for 'perpendicular.' Vectors that are linearly independent need not be orthogonal to each other [8].

[2] In mathematics the right-hand rule comes into play for the cross product of two linearly independent vectors i_n and j_n. Written as $i_n \times j_n = k_n$, the resulting k_n vector is linearly independent to the others as it is perpendicular to both and therefore normal to the plane containing them [1, 9].

These frames of reference, also known as *Cartesian coordinate systems*, are commonly used in geometry and linear algebra. For the purposes of this book, a frame of reference describing $\mathbb{R}^N$ is a Cartesian coordinate system of dimension N, or a set of N orthonormal basis vectors. In robotics, 'frame of reference' is more commonly used than 'Cartesian coordinate system.' See [1, 7, 8] for more details.

Basis vectors are labeled using various conventions, and here I use the common convention of $\mathbf{i}$, $\mathbf{j}$, and $\mathbf{k}$ to define the kinematics of a limb in a 3D Cartesian coordinate workspace.[3] A subscript n, $n+1$, etc. for a vector (as in $\mathbf{i}_n$) is used to label a specific frame of reference.[4] Even though the basis vectors of orthonormal bases are unit length, it is common to dispense with the hat because $\hat{\mathbf{i}}_n$, $\hat{\mathbf{j}}_n$, and $\hat{\mathbf{k}}_n$ is cumbersome, and because it is clear that we are using orthonormal bases.

Therefore, using this convention, any vector $\mathbf{v}_n$ in the space can be defined using frame n as

$$\mathbf{v}_n = a\,\mathbf{i}_n + b\,\mathbf{j}_n + c\,\mathbf{k}_n = \begin{pmatrix} a \\ b \\ c \end{pmatrix}_n \tag{A.11}$$

where the subscript n indicates the frame of reference being used, and the scalars a, b, and c are the projections of the vector onto each coordinate axis, as in Fig. A.1.

A.3 Rotation of Frames of Reference

Expressing vectors in a variety of frames of reference is critical to the mathematical analysis of limb kinematics and dynamics. This is because each rigid link of a limb is described mathematically by a frame of reference attached to it. From then on, you do not work with the links themselves, but only with the frames of reference that represent them. Some thorough sources for these fields are [2, 3, 10]. Here, I only present a short introduction to essential concepts from these very extensive topics.

Figure A.3 shows the example of frame 0 defined by $\mathbf{i}_0$, $\mathbf{j}_0$, and $\mathbf{k}_0$ attached to ground (which is also a rigid body), and frame 1 attached to a link that rotates about a pin joint attached to ground. Thus this is a 1 DOF kinematic chain as defined in Chap. 2. By convention, the two frames of reference share the same origin, and in their default configuration their basis vectors are aligned with each other. Then, frame 1 rotates an angle q_1 about the axis $\mathbf{k}_0$ (which is collinear with $\mathbf{k}_1$ as per the definition of default configuration). The angle q_1 is defined to be zero when both frames are aligned, and positive when rotating as per the right-hand rule. This is equivalent to saying that the link rotates in the plane of the page about the pin joint whose axis of rotation is collinear with the $\mathbf{k}_0$ and $\mathbf{k}_1$ vectors.

---

[3]Others could also be $\mathbf{x}$, $\mathbf{y}$, and $\mathbf{z}$; $\mathbf{f}_x$, $\mathbf{f}_y$, and $\mathbf{f}_z$; $\mathbf{v}_1$, $\mathbf{v}_2$, and $\mathbf{v}_3$; etc.

[4]The reason for this labeling convention will become apparent later in this Appendix when we place frames of reference sequentially along a limb, Fig. A.7.

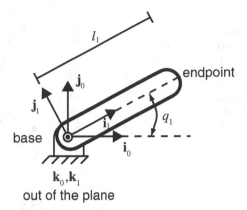

Fig. A.3 The rotation of the rigid body on the plane of the page about the kinematic DOF of a pin joint can be described as the rotation of frame 1 with respect to frame 0. Frame of reference 0 is fixed and attached to ground, and frame 1 is attached to the rigid body, and they both share the same origin. As the link rotates by an angle q_1, the location of the endpoint, Eq. A.12, does not change in frame 1, but changes in frame 0. The angle q_1 is defined to be zero when both frames are aligned, and is positive when rotating as per the right-hand rule: If your right thumb is pointing along the axis of rotation (out of the page in this case), then the fingers curl in the direction of a positive rotation

The question is, how do we know the location of the endpoint as the system rotates? Clearly, the endpoint location does not change in frame 1. It is

$$l_1\, \mathbf{i}_1 + 0\, \mathbf{j}_1 + 0\, \mathbf{k}_1 = \begin{pmatrix} l_1 \\ 0 \\ 0 \end{pmatrix}_1 \tag{A.12}$$

but what is it in the base frame 0? This will be discussed in the following section.

A.4 Rotation Matrices Using Direction Cosines

Figure A.4 shows the general form of this problem. Given a vector $\mathbf{v}$, if we know its representation in frame $n + 1$ (i.e., $\mathbf{v}_1$), then what is its representation in frame n (i.e., $\mathbf{v}_0$)? These two versions of vector $\mathbf{v}$ are a variation on Eq. A.11 written as

$$\mathbf{v}_n = a\, \mathbf{i}_n + b\, \mathbf{j}_n + c\, \mathbf{k}_n = \begin{pmatrix} a \\ b \\ c \end{pmatrix}_n \tag{A.13}$$

and

Fig. A.4 Frames of
reference n and $n + 1$ share
the same origin and are only
rotated with respect to each
other. As in Fig. A.2, vector
v exists independently of the
frames of reference, but can
be described in either frame
by its projections onto their
coordinate axes **i**, **j**, and **k**

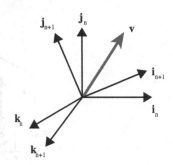

$$\mathbf{v}_{n+1} = d\,\mathbf{i}_{n+1} + e\,\mathbf{j}_{n+1} + f\,\mathbf{k}_{n+1} = \begin{pmatrix} d \\ e \\ f \end{pmatrix}_{n+1} \qquad (A.14)$$

Recall that the coefficients a, b, and c are the projections of the vector onto the basis vectors $\mathbf{i}_n$, $\mathbf{j}_n$, and $\mathbf{k}_n$; and d, e, and f are the projections of the vector defined in frame $n+1$, onto the basis vectors $\mathbf{i}_{n+1}$, $\mathbf{j}_{n+1}$, and $\mathbf{k}_{n+1}$. Therefore, the problem of expressing $\mathbf{v}_{n+1}$ in frame n reduces to the task of describing the relationships among the basis vectors of each frame of reference.

While there are multiple ways to do this, each has its advantages and disadvantages [2, 3]. An intuitive way to look at the problem is to obtain the *direction cosines* of the unit vectors of frame $n + 1$ with respect to frame n. This allows us to decompose the scalar coefficients of vector $\mathbf{v}_{n+1}$ in Eq. A.14 (i.e., d, e, and f) into the equivalent scalar coefficients for $\mathbf{v}_n$ in Eq. A.13 (i.e., a, b, and c).

The direction cosines are found by simply projecting the basis vectors of frame $n + 1$ onto the basis vectors of frame n using the *dot product* operation. Figure A.5 shows a 2D example where the projections of the $\mathbf{i}_{n+1}$ and $\mathbf{j}_{n+1}$ basis vectors onto the $\mathbf{i}_n$ and $\mathbf{j}_n$ are calculated by their respective dot products. Recall the dot product operation produces a scalar from two vectors $\mathbf{x}$ and $\mathbf{y}$,

$$\mathbf{x} \cdot \mathbf{y} = ||\mathbf{x}||\,||\mathbf{y}||\,cos(\alpha) \qquad (A.15)$$

where α is the included angle. In the case of $\mathbf{i}_{n+1} \cdot \mathbf{i}_n$ in Fig. A.5, the included angle is q. Note that because we are using unit vectors, this simplifies to[5]

$$\hat{\mathbf{x}} \cdot \hat{\mathbf{y}} = cos(\alpha) \qquad (A.16)$$

But for $\mathbf{i}_{n+1} \cdot \mathbf{j}_n$ the included angle is $(90° - q)$, which in practice is often written as $\mathbf{i}_{n+1} \cdot \mathbf{j}_n = sin(q)$ for simplicity. Equations A.27–A.29 use such simplifying substitutions.

---

[5]Such simplifications are one of the reasons why frames of reference are defined using orthonormal bases in the first place.

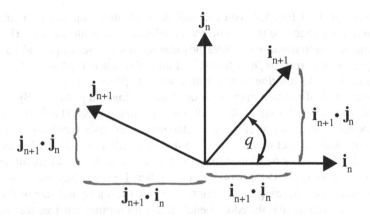

Fig. A.5 Projections of the $\mathbf{i}_{n+1}$ and $\mathbf{j}_{n+1}$ basis vectors onto the $\mathbf{i}_n$ and $\mathbf{j}_n$ basis vectors. The rotation between the two frames of reference is the angle q

If we do this systematically for every one of the basis vectors for frame $n + 1$ we obtain a way to decompose *each basis vector* $n + 1$ *onto frame* n

$$\mathbf{i}_{n+1} = (\mathbf{i}_{n+1} \cdot \mathbf{i}_n)\mathbf{i}_n + (\mathbf{i}_{n+1} \cdot \mathbf{j}_n)\mathbf{j}_n + (\mathbf{i}_{n+1} \cdot \mathbf{k}_n)\mathbf{k}_n \qquad (A.17)$$

$$\mathbf{j}_{n+1} = (\mathbf{j}_{n+1} \cdot \mathbf{i}_n)\mathbf{i}_n + (\mathbf{j}_{n+1} \cdot \mathbf{j}_n)\mathbf{j}_n + (\mathbf{j}_{n+1} \cdot \mathbf{k}_n)\mathbf{k}_n \qquad (A.18)$$

$$\mathbf{k}_{n+1} = (\mathbf{k}_{n+1} \cdot \mathbf{i}_n)\mathbf{i}_n + (\mathbf{k}_{n+1} \cdot \mathbf{j}_n)\mathbf{j}_n + (\mathbf{k}_{n+1} \cdot \mathbf{k}_n)\mathbf{k}_n \qquad (A.19)$$

These dot product operations look daunting, but they are simply the implementation of the projections as seen in Fig. A.5. It is important to work them out on your own to see for yourself. Writing this in matrix-vector form, the equality must also hold.

Defining

$$R_{n+1}^n = \begin{bmatrix} \mathbf{i}_{n+1} \cdot \mathbf{i}_n & \mathbf{i}_{n+1} \cdot \mathbf{j}_n & \mathbf{i}_{n+1} \cdot \mathbf{k}_n \\ \mathbf{j}_{n+1} \cdot \mathbf{i}_n & \mathbf{j}_{n+1} \cdot \mathbf{j}_n & \mathbf{j}_{n+1} \cdot \mathbf{k}_n \\ \mathbf{k}_{n+1} \cdot \mathbf{i}_n & \mathbf{k}_{n+1} \cdot \mathbf{j}_n & \mathbf{k}_{n+1} \cdot \mathbf{k}_n \end{bmatrix} \qquad (A.20)$$

we can write

$$\mathbf{v}_{n+1} = R_{n+1}^n \, \mathbf{v}_n \qquad (A.21)$$

The matrix R_{n+1}^n is called the *directional cosine matrix* or *direction cosine matrix*. It consists of the dot products of unit vectors of the frames n and $n + 1$.

Looking back at Fig. A.4, you can see that such direction cosine matrices are *rotation matrices* because they rotate one coordinate frame into another. Thus R^n_{n+1} rotates frame n onto frame $n+1$. Note the meaning of the superscripts and subscripts. The *superscript* is read as *from frame*, and the *subscript* is read as *to frame*. Then R^n_{n+1} is read as '*the rotation matrix from frame n to frame n + 1.*'

The structure of rotation matrices is quite interesting and powerful. By construction, the rows and columns of R^n_{n+1} all have unit magnitude. This is because they are all projections of unit vectors onto orthogonal unit basis vectors, so the sum of squares of those projections will recover that unit length as per Eq. A.6 as shown in Fig. A.5. This makes rotation matrices *orthonormal matrices*, where all rows and columns are unit vectors orthogonal to each other. Because they are orthonormal, there is the deeper meaning that rotations preserve the shape and size of the object being rotated. And there is the added benefit that orthonormal matrices are invertible, where their transpose is their inverse as follows

$$(R^n_{n+1})^{-1} = (R^n_{n+1})^T = R^{n+1}_n \tag{A.22}$$

and the last matrix reads as '*the rotation matrix from frame n+1 to frame n.*'

Thus

$$R^{n+1}_n = \begin{bmatrix} \mathbf{i}_{n+1} \cdot \mathbf{i}_n & \mathbf{j}_{n+1} \cdot \mathbf{i}_n & \mathbf{k}_{n+1} \cdot \mathbf{i}_n \\ \mathbf{i}_{n+1} \cdot \mathbf{j}_n & \mathbf{j}_{n+1} \cdot \mathbf{j}_n & \mathbf{k}_{n+1} \cdot \mathbf{j}_n \\ \mathbf{i}_{n+1} \cdot \mathbf{k}_n & \mathbf{j}_{n+1} \cdot \mathbf{k}_n & \mathbf{k}_{n+1} \cdot \mathbf{k}_n \end{bmatrix} \tag{A.23}$$

and

$$\mathbf{v}_n = R^{n+1}_n \mathbf{v}_{n+1} \tag{A.24}$$

We are finally able to address the question posed at the end of Sect. A.3 and express a same vector in two frames of reference as follows

$$\begin{pmatrix} a \\ b \\ c \end{pmatrix}_n = R^{n+1}_n \begin{pmatrix} d \\ e \\ f \end{pmatrix}_{n+1} \tag{A.25}$$

and

$$\begin{pmatrix} d \\ e \\ f \end{pmatrix}_{n+1} = R^n_{n+1} \begin{pmatrix} a \\ b \\ c \end{pmatrix}_n \tag{A.26}$$

A.5 Rotation Matrices in Practice

Given the definition of the dot product in Eq. A.15, the rotation matrices about each isolated axis are, for right-hand rule, positive rotations by an angle q,

$$R_n^{n+1}{}_{about\ \mathbf{i}} = \begin{bmatrix} 1 & 0 & 0 \\ 0 & cos(q) & -sin(q) \\ 0 & sin(q) & cos(q) \end{bmatrix} \tag{A.27}$$

$$R_n^{n+1}{}_{about\ \mathbf{j}} = \begin{bmatrix} cos(q) & 0 & sin(q) \\ 0 & 1 & 0 \\ -sin(q) & 0 & cos(q) \end{bmatrix} \tag{A.28}$$

and

$$R_n^{n+1}{}_{about\ \mathbf{k}} = \begin{bmatrix} cos(q) & -sin(q) & 0 \\ sin(q) & cos(q) & 0 \\ 0 & 0 & 1 \end{bmatrix} \tag{A.29}$$

So far I have been very explicit when using rotation matrices by using a subscript to indicate the frame in which each of the vectors is defined, as in Eq. A.24. However, it is most often clear that the super- and sub-scripts of the rotation matrix, say R_n^{n+1}, already contain that information. Therefore, unless there is ambiguity, I will not write such subscripts for the vectors to simplify notation. For example, Eq. A.25 becomes

$$\begin{pmatrix} a \\ b \\ c \end{pmatrix} = R_n^{n+1} \begin{pmatrix} d \\ e \\ f \end{pmatrix} \tag{A.30}$$

It is important that you try out these rotation matrices by doing exercises like the one shown in Fig. A.5 by hand. This way you can convince yourself of how they work and how to use them. These will become crucial to your ability to create kinematic models of limbs (Fig. A.6).

A.6 Translation and Rotation of Frames of Reference

Let us consider an extension of Fig. A.3 that includes a second link, and therefore the need to both rotate and translate frames of reference. Figure A.7 shows the example

Fig. A.6 Figure showing the isolated right-handed rotation about the collinear basis vectors $\mathbf{k_n}$ and $\mathbf{k_{n+1}}$ by an angle q that produces the rotation matrix in Eq. A.29

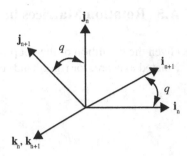

of a planar 2 DOF limb where we need to define and place a frame of reference at the limb's endpoint.

Figure A.8 shows the general case of two frames of reference displaced and rotated with respect to each other. The displacement is represented by the vector $\mathbf{p}_{n,n+1}$, which defines the vector *from* the origin of frame n *to* the origin of frame $n + 1$, *defined in frame n*. The vectors $\mathbf{v}_n$ and $\mathbf{v}_{n+1}$ both describe the same point in space, and simply express it in different frames of reference.

Equation A.31 defines how $\mathbf{v}_{n+1}$ looks when defined in frame n. The vector $\mathbf{p}_{n,n+1}$ is the translation between the frames, and $R_n^{n+1}\,\mathbf{v}_{n+1}$ is the rotation between them.

$$\mathbf{v}_n = \mathbf{p}_{n,n+1} + R_n^{n+1}\,\mathbf{v}_{n+1} \qquad\qquad\qquad \text{(A.31)}$$

Fig. A.7 Example of attaching frames of reference to two rigid bodies. The reference frame 0 fixed to ground is not shown for clarity. Notice that this requires defining frames of reference that are both rotated and translated with respect to each other

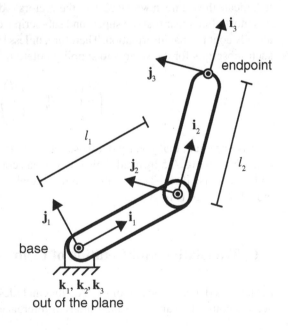

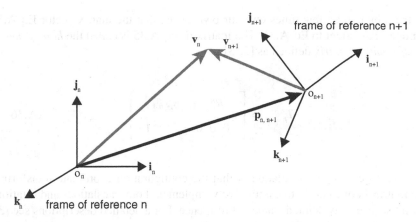

Fig. A.8 Frames of reference n and $n + 1$ displaced by vector $\mathbf{p}_{n,n+1}$. The displacement vector is expressed in frame n, and goes from the origin of frame n to the origin of frame $n + 1$. The vectors $\mathbf{v}_n$ and $\mathbf{v}_{n+1}$ describe a same point in space, but express it from different frames of reference

Explain to yourself why this equation makes sense as the addition of two vectors, both expressed in frame n.

A.7 General Case of Multiple Rigid Bodies in Series: The Homogeneous Transformation Matrix

In the general case of multiple serial linkages, as in Figs. A.7 and 2.2, it is important to note that the rotations simply concatenate—but the result of the concatenation is order-dependent and must be done systematically. Thus, if we take a vector expressed in, say, frame 5, it can be expressed in the base frame 0 as

$$v_0 = R_0^1 \ R_1^2 \ R_2^3 \ R_3^4 \ R_4^5 \ v_5 \tag{A.32}$$

$$R_0^5 = R_0^1 \ R_1^2 \ R_2^3 \ R_3^4 \ R_4^5 \tag{A.33}$$

$$v_0 = R_0^5 \ v_5 \tag{A.34}$$

But translations must all be in the same frame of references before they can be concatenated (why?). This can be done by expressing Eq. A.31 as a matrix operation, being careful to use a block structure in the matrices to ensure that the translation vector is added, and not multiplied. This can be done as follows

$$\begin{pmatrix} \mathbf{v}_n \\ 1 \end{pmatrix} = \begin{bmatrix} R_n^{n+1} & \mathbf{p}_{n,n+1} \\ 0 \ 0 \ 0 & 1 \end{bmatrix} \begin{pmatrix} \mathbf{v}_{n+1} \\ 1 \end{pmatrix} \tag{A.35}$$

Do try it by hand a few times on your own to see that the matrix vector Eq. A.35 is indeed equivalent to Eq. A.31. The matrix in Eq. A.35 is called the *homogeneous transformation matrix* defined as [2, 3]

$$T_n^{n+1} = \begin{bmatrix} R_n^{n+1} & \mathbf{p}_{n,n+1} \\ 0\ 0\ 0 & 1 \end{bmatrix} \tag{A.36}$$

An advantage of this formulation is that the configuration coordinate transformation matrices concatenate to cumulatively implement both translations and rotations among sequentially defined frames of reference. For a detailed descriptions see [3]. An example going from frame 5 to frame 0 is:

$$\mathbf{v}_0 = T_0^1\ T_1^2\ T_2^3\ T_3^4\ T_4^5\ \mathbf{v}_5 \tag{A.37}$$

$$T_0^5 = T_0^1\ T_1^2\ T_2^3\ T_3^4\ T_4^5 \tag{A.38}$$

$$\mathbf{v}_0 = T_0^5\ \mathbf{v}_5 \tag{A.39}$$

where

$$T_0^5 = \begin{bmatrix} R_0^5 & \mathbf{p}_{0,5} \\ 0\ 0\ 0 & 1 \end{bmatrix} \tag{A.40}$$

A.8 Standard Denavit–Hartenberg Convention

As a brief aside I introduce the Denavit–Hartenberg convention [11]. It is a useful and popular convention to place and describe frames of reference conveniently and concisely. You will surely encounter it in the robotics literature. It is shown for the arm example in Fig. A.9. For an in-depth discussion see [2, 11–13].

Each frame of reference (after the global frame of reference 0) is placed in a systematic manner to obey the convention and accurately define each of the 4 parameters. This convention is most often discussed as articulating frame $n - 1$ to frame n (and not from frame n to $n + 1$ as is the case for rotation and homogeneous transformation matrices).

According to the convention, we defined each axis of rotation as a z-axis about which the next distal link rotates (i.e., a revolute DOF for roboticists). To transform from 1 DOF $\mathbf{z}_{n-1}$ to the next distal DOF $\mathbf{z}_n$ via their common normal $\mathbf{x}_n$ (which points toward $\mathbf{z}_n$), one needs 4 Denavit–Hartenberg parameters (θ, d, a, α).

The 4 'standard' Denavit–Hartenberg parameters (as opposed a variant, the 'modified' Denavit–Hartenberg parameters) are specified as follows [12]:

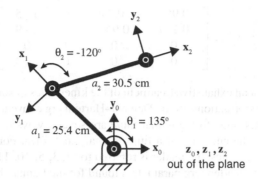

Fig. A.9 Coordinate frames assigned to planar 2 link arm with the Denavit–Hartenberg convention and corresponding Denavit–Hartenberg parameters. This example also serves to show that one can label the basis vectors of 3D frames of reference by other names, such as **x**, **y**, and **z**, which are then assigned to a specific frame of reference by their subscripts as in $\mathbf{x}_2$, $\mathbf{y}_2$, and $\mathbf{z}_2$

- Rotation of $\mathbf{x}_{n-1}$ about $\mathbf{z}_{n-1}$ by an angle θ_n
- Translation along $\mathbf{z}_{n-1}$ by d_n
- Translation along $\mathbf{x}_{n-1}$ by a_n, and
- Rotation of $\mathbf{z}_{n-1}$ about $\mathbf{x}_n$ by α_n

By convention, the DOF joint variable is θ_n for revolute joints (my q_n DOF in this book where the link length is a_n), and d_n for prismatic joints. The parameters for the arm example are shown in Fig. A.9.

Once the Denavit–Hartenberg parameters are determined, the procedure for finding the endpoint is standardized (in fact, it was developed to be implemented computationally for arbitrary systems) by means of homogeneous transformations expressed in terms of these parameters.

The 4×4 homogeneous transformation matrix is shown in Eq. A.41. The nth parameter set can be used to transform points from frame n back to frame $n - 1$.

$$
T_{n-1}^n = \begin{bmatrix}
\cos(\theta_n) & -\sin(\theta_n)\cos(\alpha_n) & \sin(\theta_n)\sin(\alpha_n) & a_n\cos(\theta_n) \\
sin(\theta_n) & \cos(\theta_n)\cos(\alpha_n) & -\cos(\theta_n)\sin(\alpha_n) & a_n\sin(\theta_n) \\
0 & \sin(\alpha_n) & \cos(\alpha_n) & d_n \\
0 & 0 & 0 & 1
\end{bmatrix} \quad \text{(A.41)}
$$

Let us take the arm shown in Fig. A.9 as a brief example. The transformation matrix T_1^2 uses the Denavit–Hartenberg parameters from frame 2 and the transformation matrix T_0^1 uses the Denavit–Hartenberg parameters from frame 1 to produce the following transformation T_0^2 matrix, from which we can directly get the coordinates of the endpoint as the last column of the homogeneous transformation matrix shown in Eq. A.42.

$$T_0^2 = T_0^1 \ T_1^2 = \begin{bmatrix} 0.966 & -0.259 & 0 & 11.5 \\ 0.259 & 0.966 & 0 & 25.9 \\ 0 & 0 & 1 & 0 \\ 0 & 0 & 0 & 1 \end{bmatrix} \qquad (A.42)$$

While this is not an exhaustive treatment of the kinematics of serial manipulators, homogeneous transformations, or the Denavit–Hartenberg convention, it suffices for the purposes of this book to know that there are several ways to find the location and orientation of the endpoint of a limb in analytical form. For a more detailed explanation of these topics, the reader is referred to [2, 3, 5, 10, 11]. In addition, an analysis of Denavit–Hartenberg parameters found for the human thumb can be seen in [12].

A.9 Exercises and Computer Code

Exercises and computer code for this section in various languages can be found at http://extras.springer.com or found by searching the World Wide Web by title and author.

References

1. G. Strang, *Introduction to Linear Algebra* (Wellesley Cambridge Press, Wellesley, 2003)
2. T. Yoshikawa, *Foundations of Robotics: Analysis and Control* (MIT Press, Cambridge, 1990)
3. R.M. Murray, Z. Li, S.S. Sastry, *A Mathematical Introduction to Robotic Manipulation* (CRC, Boca Raton, 1994)
4. Wikipedia contributors. Cartesian coordinate system. Wikipedia, The Free Encyclopedia. http://en.wikipedia.org/wiki/Cartesian_coordinate_system. Accessed 12 Feb 2015
5. F.J. Valero-Cuevas, A mathematical approach to the mechanical capabilities of limbs and fingers. Adv. Exp. Med./Biol. **629**, 619–633 (2009)
6. F.J. Valero-Cuevas, B.A. Cohn, H.F. Yngvason, E.L. Lawrence, Exploring the high-dimensional structure of muscle redundancy via subject-specific and generic musculoskeletal models. J. Biomech. **48**(11), 2887–2896 (2015)
7. Wikipedia contributors. Basis vectors. Wikipedia, The Free Encyclopedia. https://en.wikipedia.org/wiki/Basis_(linear_algebra). Accessed 12 Feb 2015
8. Wikipedia contributors. Linear independence. Wikipedia, The Free Encyclopedia. http://en.wikipedia.org/wiki/Linear_independence. Accessed 23 May 2015
9. Wikipedia contributors. Right hand rule. Wikipedia, The Free Encyclopedia. http://en.wikipedia.org/wiki/Right-hand_rule. Accessed 23 May 2015
10. T.R. Kane, D.A. Levinson, *Dynamics, Theory and Applications* (McGraw Hill, New York, 1985)
11. J. Denavit, A kinematic notation for lower-pair mechanisms based on matrices. J. Appl. Mech. **22**, 215–221 (1955)
12. V.J. Santos, F.J. Valero-Cuevas, Reported anatomical variability naturally leads to multimodal distributions of Denavit-Hartenberg parameters for the human thumb. IEEE Trans. Biomed. Eng. **53**, 155–163 (2006)
13. Wikipedia contributors. Denavit-hartenberg. Wikipedia, The Free Encyclopedia. http://en.wikipedia.org/wiki/Denavit-Hartenberg_parameters. Accessed 25 May 2015

Index

© Springer-Verlag London 2016
F.J. Valero-Cuevas, *Fundamentals of Neuromechanics*,
Biosystems & Biorobotics 8, DOI 10.1007/978-1-4471-6747-1

Printed in the United States
By Bookmasters